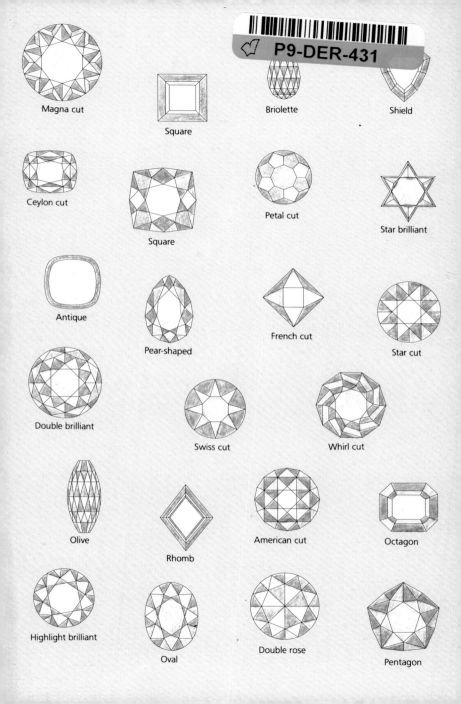

Magna cut

Square

Briolette

Shield

Ceylon cut

Square

Petal cut

Star brilliant

Antique

Pear-shaped

French cut

Star cut

Double brilliant

Swiss cut

Whirl cut

Olive

Rhomb

American cut

Octagon

Highlight brilliant

Oval

Double rose

Pentagon

GEMSTONES
of the World

*Revised &
Expanded Edition*

Walter Schumann

Sterling Publishing Co., Inc.
New York

Contents

178 Lesser-Known Gemstones

From the Preface to the First Edition

Gemstones have always fascinated mankind. In former centuries they were reserved for the ruling classes only. Today everybody can afford beautiful stones for jewelry and adornment. Precious stones for sale, especially if one includes those in so-called costume or fashion jewelry, are so numerous that it is hardly possible for the layman to survey or judge what is available. This little book has been written to help: it shows the many gems of the world in their many varieties, rough and cut, in true-to-nature color photographs. The accompanying text—always adjacent to the photographs—is designed to be of use to both the expert and the layman.

Introductory chapters on formation, properties, deposits, manufacture, synthesis, and imitations provide a survey of the world of beautiful stones. Unknown gemstones can be identified with the help of the tables at the end of the book.

I received valuable help from professional colleagues, friends, acquaintances, institutes, firms, and private persons who made stones available for illustration. My thanks are due to all of them with special thanks to Mr. Paul Ruppenthal, Idar-Oberstein. I also thank Mr. Karl Hartmann, Sobernheim, for taking the special photographs.

Preface to the Revised & Expanded Edition

Gemstones of the World has been translated into almost 20 languages. The total number of copies worldwide has passed the million mark. Again the volume was extended.

Data were brought up to date and new scientific knowledge and economic conditions were incorporated. A special chapter has been added that critically surveys the use of gemstones as cosmic-astral symbol stones and for healing purposes.

My thanks are due to many people—friends, colleagues, and amateur collectors—for their hints, suggestions, and practical help.

Walter Schumann

The Presentation of the Text

When describing gemstones, my objective has been to provide as much information as possible. Therefore, in part, smaller letter-types were used, short forms were chosen, and repetitions, which would take up space, were avoided.

The meaning of abbreviations, as well as the omission of the units of measurements, is clearly recognizable from the respective context.

Introduction

Gemstones
and Their Influence

Gems have intrigued humans for at least 10,000 years. The first known, used for making jewelry, include amethyst, rock crystal, amber, garnet, jade, jasper, coral, lapis lazuli, pearl, serpentine, emerald, and turquoise. These stones were reserved for the wealthy, and served as status symbols. Rulers sealed documents with their jewel-encrusted seals. Such treasures can now be admired at many museums and treasure-vaults.

Today, gems are worn not so much to demonstrate wealth, but rather jewelry is bought increasingly for pleasure, in appreciation of its beauty.

Certainly, also today, when purchasing a gemstone, a certain love for a special stone is part of it. Formerly, when people were less scientifically knowledgeable, gems always had an aura of mystery, something almost spiritual. That's why they were worn as amulets and talismans. Up to the present day, gemstones have sometimes been used as remedies against illnesses. They could be used in three different ways: the mere presence of the stone was sufficient to effect a cure; the gem was placed on the afflicted part of the body; or the stone was powdered and eaten.

Presently, medical science is experiencing worldwide a revival of the ideas of the Middle Ages in the use of precious stones through the doctrines of the Esoterics. The chapter on Symbol and Healing Stones (p. 247) gives more information. Gems also have an assured place in modern religion. The breastplate of the high priests of Judea was studded with four rows of gemstones. Precious stones adorn the tiara and miter of the Pope and bishops as well as the monstrances, reliquaries, and icons found in Christian churches. All major religions use precious stones, be it as decoration for tools or adorning buildings.

As a capital investment, however, of all gemstones really only diamonds are suitable. In fact, these have proven to hold on to their value, despite the travails of war or depressions in the economy.

The English Imperial State Crown with rubies, emeralds, sapphires, pearls, and more than 3000 diamonds
In the center below "Cullinan II." (or "Lesser Star of Africa") with 317.40ct. It is the fourth-largest polished diamond in the world, having 66 facets. It was cut from the largest rough diamond ever found, the "Cullinan," besides 104 other stones. The company Asscher in Amsterdam polished it (compare also "Cullinan I.," page 78, No. 3).
The large red stone above "Cullinan II." is the so-called "Black Prince's Ruby." Once thought to be a ruby, it is, in fact, a spinel, uncut, only polished, 2in (5cm) high.
The State Crown is exhibited in the Tower of London.

Terminology

In the following, important terms of the trade, used throughout the book, are explained:

Gem/Gemstone   There is no generally accepted definition for the term gem or gemstone, but they all have something special, something beautiful about them. Most gemstones are minerals (e.g., diamond), mineral aggregates (e.g., jade), or more rarely rocks (e.g., lapis lazuli). Some are organic formations (e.g., amber), and other gem materials are of synthetic origin.

There is no definite demarcation line, and woods, coal, bones, glass, and metals are all used for ornamentation. Some examples include jet (a form of coal), ivory (tusks of elephants as well as teeth of other large animals), moldavite (a glassy after-product of the striking of a meteorite), and gold nugget (more or less large gold-lumps.) Even fossils are sometimes used as ornamental material.

For some gemstones the source of specialness and beauty is the color, an unusual optical phenomenon, or the shine that makes them stand out in comparison to other stones. For other stones it is the hardness or an interesting inclusion that makes them special. Rarity also plays a role in the classification as gemstone.

Since the valued characteristics usually come into effect only through cutting and polishing, gemstones are also normally considered to be the cut stones. Cutting and polishing means refinement of what might be an otherwise insignificant raw material.

There are several hundred distinct types of gems and gem materials. The number of the variations is about double that. From time to time, new gemstones are discovered or varieties with gemstone quality are found in minerals which have been already known.

Harder stones are suitable for jewelry, whereas softer stones are often sought after by amateur collectors as well as serious lapidaries.

Gemology   Internationally, the science of gemstones is most commonly referred to as gemology.

Colored Stone   Colored stone is a trade term for all gemstones except diamonds (even those that are colorless).

Rock   With regard to gems, a rock is a natural aggregate of two or more minerals. Lapis lazuli is an example of a gem that is classified as a rock.

Semi-precious Stone   The term semi-precious has generally fallen out of use because of its derogatory meaning. Formerly, one meant with this term the less valuable and not very hard gemstones, which one opposed to the "precious" stones. "Precious" and "semi-precious" are adjectives, however, that cannot be adequately defined to distinguish between gems.

Imitation   Imitations are made to resemble natural or synthetic gem materials, completely or partially manmade. They imitate the look, color, and effect of the original substance, but they possess neither their chemical nor their physical characteristics. To these belong—strictly speaking—also those synthetic stones that do not have a counterpart in nature (for instance, fabulite or YAG). However, in the trade these are often counted as synthetic stones.

Jewel   Every individual ornamental piece is a jewel. Generally, a jewel refers to a piece of jewelry containing one or more gems set in precious metal. Sometimes, it can also refer to cut, unset gemstones.

Crystal   A crystal is a uniform body with an ordered structure, i.e., a strict order of the smallest components (the atoms, ions, or molecules) in a geometric crystal lattice. The varying structures of the lattice, together with the chemical components, are the causes of the varying physical properties of the crystals and therefore also of the gems.

Crystallography   Crystallography is the science of crystals.

Mineral   A mineral is a naturally occurring, inorganic, solid constituent of the earth's (or other celestial body's) crust. Most minerals have definite chemical compositions and crystal structures.

Mineralogy   Mineralogy is the science of minerals.

Petrography   Petrography is the descriptive science of rocks, typically utilizing a petrographic microscope or other instruments. The term petrography is often used more comprehensively to mean the same as petrology.

Petrology   Petrology is the science of the origin, history, occurrence, structure, chemical composition, and classification of rocks. Sometimes petrology is loosely used to mean the same as petrography.

Species   A mineral species is distinguished by a specific combination of chemical composition and crystal structure (e.g., diamond is carbon in a cubic structure). This combination produces the mineral's distinctive optical, physical, and chemical properties.

Stone   Popularly, stone is the collective name for all solid constituents of the earth's crust except for ice and coal. For jewelers and gem collectors, the word stone means only gemstones. For the architect, on the other hand, it means the material used for constructing buildings and streets. In the science of the earth, geology, one does not talk of stones, but of rocks and minerals.

Synthesis   Short term for a synthetic stone.

Synthetic Stone   Synthetic gemstones (in short called syntheses) are crystallized manmade products whose physical and chemical properties for the most part correspond to those of their natural gemstone counterparts.

   This term is also used by the gemstone trade for those synthetic stones which do not have a counterpart in nature (for instance, fabulite or YAG). In fact, these stones are properly termed imitations.

Variety   By a variety, one understands in the case of gemstones a modification which distinguishes itself through its look, color, or other characteristics from the actual gemstone species.

The Nomenclature of Gemstones

The oldest names for gemstones can be traced back to Oriental languages, to Greek and to Latin. Greek names especially have left their stamp on modern gem nomenclature. The meaning of old names is not always certain, especially where the first meaning of the word has been changed. Also, in antiquity, totally different stones were given the same name simply because they may have had the same color.

Original names referred to special characteristics of the stones, such as color (for instance "prase" for its green color), their place of discovery ("agate" for a river in Sicily), or mysterious powers ("amethyst" was thought to protect against drunkenness). Many mineral names, which were later also used to name gemstones, have their origin in the miners' language of the Middle Ages.

Nomenclature has been viewed scientifically only since the beginning of the modern age. Because of the discovery of many hitherto unknown minerals, new names had to be found. A principle for naming new minerals and gemstones was established which is still adhered to today. A new name is devised to refer to some special characteristic of the mineral, based on Greek or Latin, the chemical constituents, the place of occurrence, or a person's name.

Through such name-giving, not only experts are honored but also patrons or others who may or may not have any connection with mineralogy or gemology. But since everyone did not always agree when a name was given, various names for the same mineral have come into use and persist to this day.

The gemstone and jewelry trade added more of their own names, mainly to stimulate sales, producing a large number of synonyms and variety names for gemstones.

In order to correct this state of affairs, all newly discovered minerals, as well as the intended new names, must now be presented for evaluation to the Commission on New Mineral Names of the IMA (International Mineralogical Association), to which experts from all over the world belong.

Anyone who believes that he has found a new mineral or an important gemstone variety must have his first right and other legalities as well as the name-giving checked. Only then is the name of the mineral and/or the gemstone sanctioned and standardized.

Since in the gemstone trade, the danger of purposefully giving a wrong name and improper evaluation of goods is especially high, the Commission for Delivery Conditions and Quality Securing at the German Norm Commission has published in 1963/70 in the RAL A5/A5E *Guidelines for Precious Stones, Gemstones, Pearls and Corals* (for Germany) to protect against unfair competition.

Permitted definitions and the trade customs for gemstones are internationally regulated through the International Association of Jewelry, Silverwares, Diamonds, Pearls, and Stones, in short CIBJO (Confédération Internationale de la Bijouterie, Joaillerie, Orfèvrerie des Diamants, Perles, et Pierres).

In the United States, the Federal Trade Commission (FTC) *Guides for the Jewelry, Precious Metals, and Pewter Industries* serve a similar purpose.

Certainly, the aforementioned institutions have led to better communication and to more security for the gemstone buyer and seller, but such measures, naturally, cannot ensure the enforcement of an absolute guarantee for genuineness.

False and Misleading Names of Some Gemstones

False Gemstone Name	Preferred Gemnological Name
Adelaide ruby	Almandite
African emerald	Green fluorite
Alaska diamond	Rock crystal (quartz)
American jade	Green idocrase

American ruby	Pyrope or almandite (garnet) or rose quartz
Arizona ruby	Pyrope (garnet)
Arizona spinel	Red or green garnet
Arkansas diamond	Rock crystal (quartz)
Balas ruby	Red spinel
Blue alexandrite	Color-change sapphire
Blue moonstone	Artificially blue-tinted chalcedony
Bohemian chrysolite	Moldavite (natural glass)
Bohemian diamond	Rock crystal (quartz)
Bohemian ruby	Pyrope (garnet) or rose quartz
Brazilian aquamarine	Bluegreen topaz
Brazilian ruby	Red or pink topaz
Brazilian sapphire	Blue tourmaline
Californian ruby	Hessonite (grossular garnet)
Candy spinel	Almandite (garnet)
Cape-chrysolite	Green prehnite
Cape-ruby	Pyrope (garnet)
Ceylon diamond	Colorless zircon
Ceylon opal	Opal-like glimmery moonstone
Copper lapis	Azurite
German diamond	Rock crystal (quartz)
German lapis	Artificially blue-tinted jasper (chalcedony)
Gold topaz	Citrine (quartz)
Indian jade	Aventurine (quartz)
King's topaz	Yellow sapphire
Korean jade	Serpentine
Lithia amethyst	Kunzite (spodumene)
Lithia emerald	Hiddenite (spodumene)
Madeira topaz	Citrine (quartz)
Marmarosch diamond	Rock crystal (quartz)
Matura diamond	Colorless fired zircon
Mexican diamond	Rock crystal (quartz)
Mexican jade	Artificially tinted green marble
Montana ruby	Red garnet
Oriental amethyst	Violet sapphire
Oriental hyacinth	Pink sapphire
Oriental topaz	Yellow sapphire
Palmyra topaz	Brown synthetic sapphire
Salmanca topaz	Citrine (quartz)
Saxon chrysolite	Greenish-yellow topaz
Saxon diamond	Colorless topaz
Serra topaz	Citrine (quartz)
Siamese aquamarine	Blue zircon
Siberian chrysolite	Demantoid (garnet)
Siberian ruby	Red tourmaline
Simili diamond	Glass imitation
Slave-diamond	Colorless topaz
Smoky topaz	Smoky quartz
Spanish topaz	Citrine (quartz)
Strass diamond	Glass imitation
Transvaal jade	Green hydrogrossular garnet
Ural sapphire	Blue tourmaline
Viennese turquoise	Artificially blue-tinted argillaceous earth

Origin and Structure of Gemstones

Since, with few exceptions, most gemstones are minerals, it is the origin and structure of minerals that concerns the gemologist. The formation of the nonmineral gemstones (for instance, amber, coral, and pearl) will be dealt with in more detail when they are described.

Origin Minerals can be formed in various ways. Some crystallize from molten magma and gases of the earth's interior or from volcanic lava streams that reach the earth's surface (igneous or magmatic minerals). Others crystallize from hydrous solutions or grow with the help of organisms on or near the earth's surface (sedimentary minerals). Lastly, new minerals are formed by recrystallization of existing minerals under great pressure and high temperatures in the lower regions of the earth's crust (metamorphic minerals).

The chemical composition of the minerals is shown by a formula. Impurities are not included in this formula, even where they cause the color of the stone.

Formations of Crystals Nearly all minerals grow in certain crystal forms, i.e., they are homogeneous bodies with a regular lattice of atoms, ions, or molecules. They are geometrically arranged and their outer shapes are limited by flat surfaces (in the ideal case), resulting in crystal faces.

Most crystals are small, sometimes even microscopically small, but there are also some giant specimens. In general, the smallest minerals (due to their tiny size) as well as the giant minerals (due to their inclusions, impurities, or uneven growth marks) are unsuitable as gems.

The mineral's chemical composition and inner structure, the lattice, determine the physical properties of the crystal (its outer shape, hardness, cleavage, type of fracture, specific gravity) and also its optical properties.

Crystal lattice of diamond.

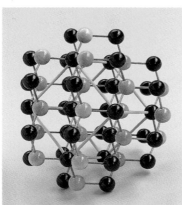

Crystal lattice of quartz.

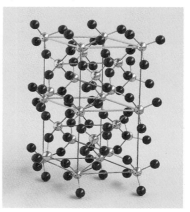

Rubellite, Madagascar (2/3 natural size).

Most crystals are not regularly shaped, but have an irregular form, because some crystal faces have developed better at the cost of others; however, the angle between the faces always remains constant. When individual single crystals occur in combination with other crystal forms, e.g., hexahedron with octahedron, the identification of a mineral on the basis of crystal shape can be extraordinarily complicated.

The arrangement of faces preferred by a mineral is called the "habit"; for instance, pyrite is often found in the shape of a pentagon dodecahedron, garnet, as a rhomb-dodecahedron. The habit of a crystal also refers to its type and can be tabular, acicular, foliated, columnar, or compact. For the benefit of laymen, the technical terms, habit, and form are sometimes called structure. Commonly minerals adopt the crystal form of another—usually through replacing it. These occurrences are called pseudomorphs.

Where two or more crystals are intergrown according to certain laws, one speaks of twins, triplets, or quadruplets. Depending on whether the individual crystals are grown together or intergrown, one speaks of contact twins or penetration twins.

Apart from twinning which adheres to certain laws, many crystals are irregularly intergrown into aggregates. Depending on the growth process, filiform (wire-like), fibrous, radial-shaped, leaf-like, shell-like, scaly, or grainy aggregates are formed. According to miners' lingo, a mineral aggregate with free-standing individual crystals is called "step."

Well-developed, characteristic minerals are formed as druses on the inner walls of rock openings (geodes); these are mainly round hollows created by gas bubbles in magmatic rocks or spaces where organic material has been removed in sedimentary rocks.

Crystal systems   In crystallography, crystals are divided into seven systems. The distinction is made according to crystal axes and the angles at which the axes intersect. On page 17, the crystal systems are depicted with some typical crystal formations.

Cubic system   (also called the isometric system)   All three axes have the same length and intersect at right angles. Typical crystal shapes are the cube, octahedron, rhombic dodecahedron, icosi-tetrahedron, and hexacisochedron.

Tetragonal system   The three axes intersect at right angles, two are of the same length and are in the same plane, while the main axis is either longer or shorter. Typical crystal shapes are four-sided prisms and pyramids, trapezohedrons and eight-sided pyramids as well as double pyramids.

Hexagonal system   Three of the four axes are in one plane, are of the same length, and intersect each other at angles of 60 degrees. The fourth axis, which is a different length, is at right angles to the others. Typical crystal shapes are hexagonal prisms and pyramids, as well as twelve-sided pyramids and double pyramids.

Trigonal system   (rhombohedral system)   Axes and angles are similar to the preceding system, therefore the two systems are often combined as hexagonal. The difference is one of symmetry. In the case of the hexagonal system, the cross section of the prism base is six-sided; in the trigonal system, it is three-sided. The six-sided hexagonal shape is formed by a cutting-off process of the corners of the triangles. Typical crystal forms of the trigonal system are three-sided prisms and pyramids, rhombohedra, and scalenohedra.

Orthorhombic system   (rhombic system)   Three axes of different lengths are at right angles to each other. Typical crystal shapes are basal pinacoids, rhombic prisms, and pyramids as well as rhombic double pyramids.

Monoclinic system   The three axes are each of different lengths, two are at right angles to each other, and the third one is inclined. Typical crystal forms are basal pinacoids and prisms with inclined end faces.

Triclinic system   All three axes are of different lengths and inclined to each other. Typical crystal forms are paired faces.

In the table on page 18, minerals from which many gemstone varieties come are allocated to their respective crystal systems.

Horizontal gemstone microscope. In the center of the apparatus is a box with a red gemstone, which is held by a movable claw holder.

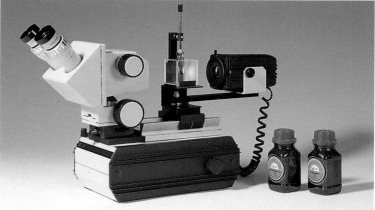

Crystal Systems

cubic

tetragonal

hexagonal
and
trigonal

orthorhombic

monoclinic

triclinic

17

Selected Gemstones
Ordered by Crystal System

cubic

Almandite
Analcite
Andradite
Cubic zirconia (CZ)
Cuprite
Demantoid
Diamond
Fluorite
Gahnite
Garnet
GGG
Gold
Grossularite
Haüynite
Hessonite
Lazurite
Magnetite
Melanite
Periclase
Pollucite
Pyrite
Pyrope
Senarmontite
Silver
Sodalite
Spessartite
Sphalerite
Spinel
Strontium titanate
Topazolite
Tsavorite
Uvarovite
YAG

tetragonal

Anatase
Apophyllite
Boleite
Carletonite
Cassiterite
Chalcopyrite
Idocrase
Leucite
Melinophane
Phosgenite
Pyrolusite
Rutile
Scapolite
Scheelite
Tugtupite
Wardite
Wulfenite
Zircon

hexagonal

Apatite
Aquamarine
Benitoite
Beryl
Cancrinite
Catapleiite
Emerald
Goshenite
Greenockite
Heliodor
Jeremejevite
Milarite
Morganite
Nepheline
Painite
Sugilite
Taaffeite
Thaumasite
Zincite

trigonal

Agate
Amethyst
Aventurine
Calcite
Carnelian
Chalcedony
Chrysoprase
Cinnabar
Citrine
Corundum
Davidite
Dioptase
Dolomite
Eudialyte
Friedelite
Gaspeite
Hematite
Jasper
Magnesite
Parisite
Phenakite
Prasiolite
Proustite
Pyrargyrite
Quartz
Rhodochrosite
Rock crystal
Rose quartz
Ruby
Sapphire
Siderite
Simpsonite
Smithsonite
Smoky quartz
Stichtite
Tiger's-eye
Tourmaline
Willemite

orthorhombic

Adamite
Alexandrite
Andalusite
Anglesite
Anhydrite
Aragonite
Barite
Boracite
Celestite
Cerussite
Chrysoberyl
Danburite
Descloizite
Diaspore
Dumortierite
Enstatite
Hambergite
Hemimorphite
Hypersthene
Iolite
Kornerupine
Lithiophilite
Manganotantalite
Meerschaum
Mesolite
Natrolite
Peridot
Prehnite
Purpurite
Sulfur
Sinhalite
Strontianite
Tanzanite
Tantalite
Thulite
Topaz
Triphylite
Variscite
Witherite
Zoisite
Zektzerite

monoclinic

Aegerine-augite
Azurite
Barytocalcite
Beryllonite
Brazilianite
Charoite
Chrysocolla
Clinohumite
Clinozoisite
Colemanite
Crocoite
Diopside
Eosphorite
Epidote
Euclase

Gaylussite
Gypsum
Hiddenite
Hornblende
Howlite
Jadeite
Kunzite
Lazulite
Legrandite
Malachite
Moonstone
Muscovite
Nephrite
Neptunite
Orthoclase
Petalite
Phosphopyllite
Sapphirine
Serpentine
Sphene
Spodumene
Staurolite
Talc
Tremolite
Vivianite
Vlasovite
Whewellite
Yugawaralite

triclinic

Amazonite
Amblygonite
Andesine
Aventurine feldspar
Axinite
Kurnakovite
Kyanite
Labradorite
Microcline
Montebrasite
Oligoclase
Pectolite
Rhodonite
Sanidine
Turquoise
Ulexite

amorphous

Amber
Ekanite
Moldavite
Obsidian
Opal

Properties of Gemstones

Specific knowledge about the most important properties of gemstones are of inestimable value to the gemstone cutter and the setter, as well as to the wearer of the jewelry and the collector. Only with proper knowledge can one correctly work the gemstone, make use of it, and take care of it.

Hardness

In the case of minerals and gemstones, hardness refers first to scratch hardness, then to cutting resistance.

Scratch Hardness

The Viennese mineralogist Friedrich Mohs (1773–1839) introduced the term *scratch hardness*. He defined the scratch hardness as the resistance of a mineral when scratched with a pointed testing object.

Mohs set up a comparison scale using ten minerals of different degrees of hardness (Mohs' hardness scale), which is still widely in use. Number 1 is the softest, number 10 the hardest degree. Each mineral in the series scratches the previous one with the lesser hardness and is scratched by the one which follows after. Minerals of the same hardness will also scratch each other. In practice, the hardness grades have also been subdivided into half degrees. All minerals and gemstones known to us today are allocated to Mohs' hardness scale. (See table, page 21.)

Gemstones of the scratch hardness (Mohs' hardness) 1 and 2 are considered soft, those of the degrees 3 to 5 medium hard, and those over 6 hard. Formerly, one spoke also of gemstone hardness in regards to the steps 8 to 10. This is no longer in use, since there are valuable gemstones that do not have these high Mohs' hardness values.

The luster and polish of gemstones of the Mohs' hardness below 7 can be damaged by dust, as this may contain small particles of quartz (Mohs' hardness 7). Through the scratching of this quartz dust, the stones become dull in the course of time. Such stones must be carefully handled when being worn and stored.

Relative and Absolute Hardness Scale

Scratch hardness (Mohs)	Mineral used for comparison	Simple hardness tester	Cutting resistance (Rosiwal)
1	Talc	Can be scratched with fingernail	0.03
2	Gypsum	Can be scratched with fingernail	1.25
3	Calcite	Can be scratched with copper coin	4.5
4	Fluorite	Easily scratched with knife	5.0
5	Apatite	Can be scratched with knife	6.5
6	Orthoclase	Can be scratched with steel file	37
7	Quartz	Scratches window glass	120
8	Topaz		175
9	Corundum		1,000
10	Diamond		140,000

The Mohs' hardness scale is a relative hardness scale. It only shows which gemstone is harder than another one. Nothing is said about increase of hardness within the scale. That is possible only with absolute hardness scales, for example, cutting resistance. (See table, page 19.)

Hardness Test　Formerly, when the optical examination methods had not been developed to such a degree as they are today, the scratch-hardness test played a larger role for the determination of gemstones. Now, the scratch-hardness test is used only very rarely by gemologists. It is too inaccurate for a precise testing of hardness, and there is the danger of hurting the gemstone.

In the trade, test pieces and scratch tools for the testing of hardness can be bought. When doing the scratch test, one must make sure that the examination is done only with sharp-edged objects on fresh, nondecomposed crystal or cut surfaces. Corrugated or foliated formations feign a lesser hardness.

Always begin with softer test materials in order not to damage the gemstone unnecessarily. If possible, do not scratch transparent cut stones at all. With translucent or opaque gemstones test only at an unnoticeable spot, or on the underside.

Cutting Resistance

For the gemstone cutter, naturally the hardness of a stone, when cutting, plays an important role. There are also gemstones which are of a different hardness on different crystal faces and in different directions. For the stone collector, small differences in hardness are of lesser importance. In kyanite (compare drawing below and page 196), for example, the Mohs' hardness along the stem-like crystals is 4½, but across it is 6 to 7. There are also important differences in hardness in diamond on the crystal faces (see drawing below). (For more about the cutting of diamond, see page 64.)

For the gemstone cutter, it would be helpful to have absolute values regarding the cutting hardness of gemstones. Unfortunately, there are hardly any useful numbers available. The cutter must rather discover for himself in practice and rely on his experience.

It is real art to cut softer gemstones, which only a few specialists master. If the crystal faces of a stone, in addition, are of different degrees of hardness, it takes a lot of skill to form sharp and even edges on these stones.

When polishing gemstones, the hardness is of utmost importance, because harder gemstones take a polish better than softer stones.

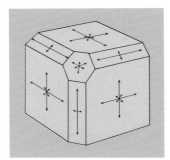

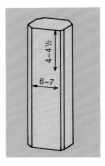

Left: Differences in hardness in diamond. The shorter the arrow, the larger the cutting hardness in this direction (according to E.M. and J. Wilks).

Right: Differences in hardness in kyanite. Clearly different Mohs' hardnesses in longitudinal and transverse directions.

Selected Gemstones Ordered by Mohs' Hardness

Diamond	10	Tanzanite	6½–7	Obsidian	5–5½
Corundum	9	Tiger's-eye	6½–7	Sphene	5–5½
Ruby	9	GGG	6½	Wolframite	5–5½
Sapphire	9	Idocrase	6½	Apatite	5
Alexandrite	8½	Epidote	6–7	Dioptase	5
Chrysoberyl	8½	Cassiterite	6–7	Eosphorite	5
Cubic zirconia	8½	Pyrolusite	6–7	Glass	5
Djevalite	8–8½	Amazonite	6–6½	Hemimorphite	5
Spinel	8	Andesine	6–6½	Smithsonite	5
Taaffeite	8	Aventurine Fsp.	6–6½	Apophylite	4½–5
Topaz	8	Benitoite	6–6½	Gaspeite	4½–5
YAG	8	Labradorite	6–6½	Scheelite	4½–5
Aquamarine	7½–8	Moonstone	6–6½	Colemanite	4½
Beryl	7½–8	Nephrite	6–6½	Kyanite	4–7
Emerald	7½–8	Orthoclase	6–6½	Variscite	4–5
Gahnite	7½–8	Petalite	6–6½	Zincite	4–5
Phenakite	7½–8	Prehnite	6–6½	Purpurite	4–4½
Andalusite	7½	Pyrite	6–6½	Ammonite	4
Euclase	7½	Rutile	6–6½	Barytocalcite	4
Hambergite	7½	Sugilite	6–6½	Fluorite	4
Sapphirine	7½	Tantalite	6–6½	Rhodochrosite	4
Dumortierite	7½–8½	Zoisite	6–6½	Magnesite	3½–4½
Boracite	7–7½	Amblygonite	6	Siderite	3½–4½
Danburite	7–7½	Sanidine	6	Aragonite	3½–4
Iolite	7–7½	Hematite	5½–6½	Azurite	3½–4
Simpsonite	7–7½	Magnetite	5½–6½	Chalcopyrite	3½–4
Tourmaline	7–7½	Opal	5½–6½	Cuprite	3½–4
Amethyst	7	Rhodonite	5½–6½	Dolomite	3½–4
Aventurine	7	Actinolite	5½–6	Malachite	3½–4
Rock crystal	7	Anatase	5½–6	Sphalerite	3½–4
Citrine	7	Beryllonite	5½–6	Anhydrite	3½
Prasiolite	7	Haüynite	5½–6	Coral	3–4
Quartz	7	Leucite	5½–6	Barite	3–3½
Smoky quartz	7	Periclase	5½–6	Cerussite	3–3½
Rose quartz	7	Scapolite	5½–6	Celestite	3–3½
Almandite	6½–7½	Sodalite	5½–6	Howlite	3–3½
Andradite	6½–7½	Tugtupite	5½–6	Witherite	3–3½
Demantoid	6½–7½	Brazilianite	5½	Calciate	3
Garnet	6½–7½	Enstatite	5½	Kurnakovite	3
Grossularite	6½–7½	Linobate	5½	Wulfenite	3
Hessonite	6½–7½	Moldavite	5½	Serpentine	2½–5½
Pyrope	6½–7½	Smaragdite	5½	Pearl	2½–4½
Spessartite	6½–7½	Willemite	5½	Gaylussite	2½–3
Uvarovite	6½–7½	Cancrinite	5–6	Gold	2½–3
Zircon	6½–7½	Catapleiite	5–6	Crocoite	2½–3
Agate	6½–7	Charoite	5–6	Silver	2½–3
Axinite	6½–7	Diopside	5–6	Proustite	2–2½
Chalcedony	6½–7	Hypersthene	5–6	Chrysocolla	2–4
Chrysopase	6½–7	Lapis Lazuli	5–6	Phosgenite	2–3
Diaspore	6½–7	Lazulite	5–6	Amber	2–2½
Hiddenite	6½–7	Strontium		Cinnabar	2–2½
Jadeite	6½–7	titanate	5–6	Meerschaum	2–2½
Jasper	6½–7	Tremolite	5–6	Ulexite	2–2½
Kornerupine	6½–7	Turquoise	5–6	Gypsum	2
Kunzite	6½–7	Analcite	5–5½	Sulfur	1½–2½
Peridot	6½–7	Datolite	5–5½	Stichtite	1½–2½
Pollucite	6½–7	Melinophane	5–5½	Vivianite	1½–2
Sinhalite	6½–7	Natrolite	5–5½	Talc	1

Cleavage
and Fracture

Many gemstones can be split along certain flat planes, which the expert calls *cleavage*. Cleavage is related to the lattice of the crystal—the cohesive property of the atoms.

Depending on the ease with which a crystal can be cleaved, one differentiates between a perfect (euclase), a good (sphene), and an imperfect cleavage (peridot). Some gemstones cannot be cleaved at all (quartz). One then says, they do not have cleavage. A loosening of contact twins is not called cleavage, but separation or parting.

Lapidaries and stone setters must take account of the cleavage. Often a small tap or too much pressure when testing for Mohs' hardness is sufficient to split the stone. When soldering, the temperature can cause fissures along the cleavage planes, where eventually the gem may break completely along this line. In gemstones with perfect cleavage the facets must be transverse to the cleavage planes, otherwise the stone will be more vunerable to breaking. Piercings should, if possible, be done vertically to the cleavage surfaces.

Cleavage is used to divide large gem crystals or remove faulty pieces. The largest diamond of gem quality ever found, the Cullinan of 3106cts, was cleaved in 1908 into three large pieces which were then cleaved into numerous smaller pieces. Today, small pieces are usually sawn in order to avoid unwanted cleavages, and to make the best use of the shape of the stone.

The breaking of a gemstone with a blow producing irregular surfaces is called *fracture*. It can be conchoidal (shell-like), uneven, smooth, fibrous, splintery, or grainy. Sometimes, the type of fracture helps to identify a mineral. Conchoidal fracture is, for instance, characteristic for all quartz and glass-like minerals.

Density
and Specific Gravity

In general scientific usage the measurement *specific gravity*, which indicates the ratio of the weight of a specific material to the weight of the same volume of water, is now replaced by the term *density*, which is expressed typically as grams per cubic centimeter (g/cm^3).

Weight is in fact not a constant attribute. It depends on the magnitude of the gravity at the respective location where it is measured. But for the determination of the weight of gemstones, this does not matter, since the measurement is always done under the same gravitation conditions. A more obvious factor affecting weight is an object's size.

The density is a property independent of location and size. It is defined as weight per volume represented in g/cm^3 and/or kg/m^3.

In practice, there is not a significant distinction between using the term density and specific gravity, since nothing is really changed in actual measurement. Nevertheless, in the following, the term *density* is used instead of *specific gravity*.

In regards to the "heaviness" of a gemstone, we will continue to speak of "weight" and not of the "density" of a gemstone, since that corresponds to the practice of the gemstone trade.

The density of gemstones varies between 1 and 8. Values under 2 are considered light (e.g., amber about 1), those from 2 to 4 normal (quartz 2.6), and those over 4 are considered heavy (casserite around 7). The more valuable gemstones (such as diamond, ruby, and sapphire) have densities that are much greater than the common rock-forming minerals, especially quartz and feldspar. Such heavy minerals are deposited in flowing waters before the sands, which are rich in quartz, and they form the so-called placer deposits (compare with page 53).

Determination of Density

In identifying gemstones, a determination of the density can be useful. But for specialists, optical procedures for the determination of gemstones are more common. These procedures require expensive instruments, however.

Two methods have proven a success for the determination of the density of gemstones: the buoyancy method with the help of a hydrostatic balance and the suspension, or heavy liquid, method. The first is time consuming, but inexpensive. It is also inaccurate with small stones. Even though the second method is a little bit more expensive, it produces good results in a short time, especially with lots of unknown gems. However, the immersion liquids are quite hazardous.

Hydrostatic Balance   The measuring procedure with a hydrostatic scale works on Archimedes' Principle of buoyancy. The volume of the unknown gem is determined. The density is then easily worked out. A hydrostatic balance can be constructed by anybody (see illustration below). The beginner can adapt letter scales. Anyone more advanced should use a precision balance as used by a chemist or pharmacist. The object to be tested is first weighed in air (in the pan under the bridge) and then in water (in the net in the beaker).

Hydrostatic balance for determining gemstone volume.

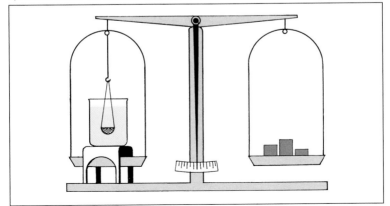

The difference in weight corresponds to the weight of the water that is displaced, and therefore equal to the volume of the gemstone. It is possible to determine the density with this method to one, and with practice, to two decimal places. It is important to ensure that the stone is not in contact with a foreign substance, that it is not set, and that, when weighed in air, it is dry. It is also necessary to take into account the weight of the net used to suspend the gem in the beaker.

Example:

Substance (weight) in air	5.2 g	Density = $\dfrac{\text{Weight in air}}{\text{Volume}}$ =	$\dfrac{5.2}{1.9}$	
Substance (weight) in water	3.3 g			
Difference = Volume	1.9 cm³		=	2.7 g/cm³

Suspension Method   This method rests on the idea that an object will float in a liquid of higher density, sink in a liquid of lower density, and remain suspended in a liquid of the same density. To perform the test, a set of liquids with known high densities (heavy liquids) is normally used. One set sold commercially includes liquids with densities of 2.57, 2.62, 2.67, 3.05, and 3.32. This range and others similar are adequate for testing most gemstones. Certain liquids are also useful in specific situations (e.g., 2.67 liquid for distinguishing between natural and synthetic emerald).

To determine the density, the gem is placed in one of the heavy liquids and released about midway in the container. If the gem remains suspended, its density is the same as the liquid's and no further testing is needed. If the gem floats, the test is repeated using the liquid of next lower density. If the gem sinks, the liquid of next higher density is used. By this process of elimination, the liquids with densities nearest to the gem's are identified, then the gem's density is estimated by its rate of rising or sinking in those liquids. For example, if the gem rises slowly in 3.05 liquid but sinks rapidly in 2.67 liquid, its density can be estimated as 3.00. The gem's rate of rising or sinking can also be compared to that of an indicator stone of known density.

This method is especially useful for testing small stones and large numbers of gemstones. However, it provides only an approximate estimate of density. Most heavy liquids are toxic as well. Therefore, one must follow proper precautions when using them. Do not breath in the vapors, and do not eat while you are working. Heavy liquids can also damage organic or porous gemstones such as amber, ivory, lapis lazuli, pearl, shell, and turquoise.

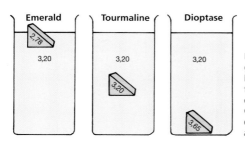

Determining the density of a gemstone with the help of heavy liquids. Lighter stones float at the surface; heavier ones sink to the bottom. Gemstones with the same density as that of the liquid are suspended in the liquid.

Selected Gemstones Ordered by Density

Gemstone	Density	Gemstone	Density	Gemstone	Density
Gold	15.5–19.3	Sinhalite	3.46–3.50	Amethyst	2.65
Silver	9.6–12.0	Rhodochrosite	3.45–3.70	Rock crystal	2.65
Cinnabar	8.0–8.2	Rhodonite	3.40–3.74	Citrine	2.65
Wolframite	7.1–7.6	Sapphirine	3.40–3.58	Prasiolite	2.65
GGG	7.05	Aegirine–augite	3.40–3.55	Quartz	2.65
Cassiterite	6.7–7.1	Hypersthene	3.4–3.5	Smoky quartz	2.65
Wulfenite	6.50–7.00	Tanzanite	3.35	Rose quartz	2.65
Cerussite	6.46–6.57	Triphylite	3.34–3.58	Aventurine	2.64–2.69
Phosgenite	6.13	Idocrase	3.32–3.47	Aventurine Fsp.	2.62–2.65
Simpsonite	5.92–6.84	Epidote	3.3–3.5	Pearl	2.60–2.85
Scheelite	5.9–6.3	Hemimorphite	3.30–3.50	Coral	2.60–2.70
Crocoite	5.9–6.1	Diaspore	3.30–3.39	Agate	2.60–2.64
Cuprite	5.85–6.15	Jadeite	3.30–3.38	Petrified wood	2.58–2.91
Zincite	5.66	Peridot	3.28–3.48	Jasper	2.58–2.91
Proustite	5.51–5.64	Dioptase	3.28–3.35	Iolite	2.58–2.66
Cubic zirconia	5.5–5.9	Kornerupine	3.27–3.45	Chalcedony	2.58–2.64
Tantalite	5.18–8.20	Dumortierite	3.26–3.41	Chrysoprase	2.58–2.64
Hematite	5.12–5.28	Axinite	3.26–3.36	Moss agate	2.58–2.64
Strontium titanate	5.11–5.15	Malachite	3.25–4.10	Tiger's-eye	2.58–2.64
Pyrite	5.00–5.20	Smaragdite	3.24–3.50	Scapolite	2.57–2.74
Linobate	4.64–4.66	Diopside	3.22–3.38	Moonstone	2.56–2.59
YAG	4.55	Purpurite	3.2–3.4	Amazonite	2.56–2.58
Pyrolusite	4.5–5.0	Enstatite	3.20–3.30	Talc	2.55–2.80
Davidite	4.5	Apatite	3.16–3.23	Charoite	2.54–2.78
Barite	4.43–4.46	Hiddenite	3.15–3.21	Lapis lazuli	2.50–3.00
Witherite	4.27–4.35	Kunzite	3.15–3.21	Howlite	2.45–2.58
Rutile	4.20–4.30	Euclase	3.10	Leucite	2.45–2.50
Spessartite	4.12–4.18	Andalusite	3.05–3.20	Carletonite	2.45
Chalcopyrite	4.10–4.30	Lazulite	3.04–3.14	Serpentine	2.44–2.62
Smithsonite	4.00–4.65	Actinolite	3.03–3.07	Variscite	2.42–2.58
Gahnite	4.00–4.62	Amblygonite	3.01–3.11	Cancrinite	2.42–2.51
Legrandite	3.98–4.04	Fluorite	3.00–3.25	Hauynite	2.4–2.5
Ruby	3.97–4.05	Melinophane	3.00–3.03	Colemanite	2.40–2.42
Celestite	3.97–4.00	Montebrasite	2.98–3.11	Petalite	2.40
Sapphire	3.95–4.03	Brazilianite	2.98–2.99	Tugtupite	2.36–2.57
Zircon	3.93–4.73	Danburite	2.97–3.03	Obsidian	2.35–2.60
Almandite	3.93–4.30	Tremolite	2.95–3.07	Hambergite	2.35
Sphalerite	3.90–4.10	Phenakite	2.95–2.97	Moldavite	2.32–2.38
Willemite	3.89–4.18	Aragonite	2.94	Turquoise	2.31–2.84
Siderite	3.83–3.96	Hornblende	2.9–3.4	Apophyllite	2.30–2.50
Uvarovite	3.77	Nephrite	2.90–3.03	Mesolite	2.26–2.40
Andradite	3.7–4.1	Datolite	2.90–3.00	Analcite	2.22–2.29
Azurite	3.7–3.9	Anhydrite	2.90–2.98	Natrolite	2.20–2.26
Periclase	3.7–3.9	Pollucite	2.85–2.94	Yugawaralite	2.19–2.23
Chrysoberyl	3.70–3.78	Tourmaline	2.82–3.32	Stichtite	2.16–2.18
Barytocalcite	3.66	Prehnite	2.82–2.94	Sodalite	2.14–2.40
Staurolite	3.65–3.77	Dolomite	2.80–2.95	Sulfur	2.05–2.08
Benitoite	3.64–3.68	Beryllonite	2.80–2.87	Chrysocolla	2.00–2.40
Strontianite	3.63–3.79	Muscovite	2.78–2.88	Meerschaum	2.0–2.1
Pyrope	3.62–3.87	Sugilite	2.76–2.80	Gaylussite	1.99
Taaffeite	3.60–3.62	Ammonite	2.75–2.80	Opal	1.98–2.50
Grossularite	3.57–3.73	Catapleiite	2.72	Thaumasite	1.91
Spinel	3.54–3.63	Calcite	2.69–2.71	Kurnakovite	1.86
Kyanite	3.53–3.70	Aquamarine	2.68–2.74	Ivory	1.7–2.0
Sphene	3.52–3.54	Emerald	2.67–2.78	Ulexite	1.65–1.95
Diamond	3.50–3.53	Precious beryl	2.66–2.87	Jet	1.19–1.35
Topaz	3.49–3.57	Labradorite	2.65–2.75	Amber	1.05–1.09
		Andesine	2.65–2.69		

Weights Used
in the Gem Trade

In the international gem trade, the carat, gram, grain, and momme are used as units of weight.

Carat   The weight used in the gem trade since antiquity. The name is derived from the seed *kuara* of the African Coraltree or from the kernel (Greek *keration*) of the Carob bean.

Since 1907 Europe and America, followed by other countries, have adopted the metric carat (mct) of 200 mg or 0.2 g. Local carat weights have differed historically from location to location, varying between 188 and 213 mg.

The carat is subdivided into fractions (e.g., 1/10ct) or decimals (e.g., 1.25ct) up to two decimal places. Small diamonds are weighed in *points*, which are 1/100 carat (0.01ct). The table below illustrates diameter and corresponding carat weight for diamonds cut in the modern brilliant cut (see page 81). Gems with different density and different cuts obviously have different diameters.

The price of a gemstone is usually indicated in the gem trade "per carat." By calculating the actual weight, one receives the price per piece. When selling to the final buyer, though, usually the total price is given. The carat price often progressively increases with the size of the gemstone in very small increments. When a one-carat piece, for example, costs $750, then a two-carat-piece is not necessarily worth $1500, but maybe $3000 or even more.

The carat weight of gems is not to be confused with the karat used by the goldsmith. In the case of gold the karat is not a weight measure at all but rather a measure of quality. The higher the karatage, the higher the content of gold in the piece of jewelry. The weight can be variable.

Gram   The weight measure used in the trade for less precious gemstones and especially for rough stones.

Grain   Formerly, the weight measure for pearls. It corresponds to 0.05g or 1/4ct. Today, it is increasingly substituted by use of the carat.

Momme   The old Japanese measure of *momme* (=3.75g=18.75ct) for cultured pearls is hardly used anymore except by wholesale distributors.

Diameters and Weights of Brilliant Diamonds

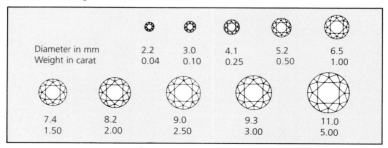

Diameter in mm	2.2	3.0	4.1	5.2	6.5
Weight in carat	0.04	0.10	0.25	0.50	1.00
	7.4	8.2	9.0	9.3	11.0
	1.50	2.00	2.50	3.00	5.00

Optical Properties

Of all the various properties of gemstones, the optical characteristics are of unsurpassed importance. They produce color and luster, fire and luminescence, play of light, and schiller (iridescence). In the examination of gems, nowadays, there is more and more concentration on the optical effects.

Color

Color is the most important characteristic of gems. In the case of most stones, it is not diagnostic in identification, because many have the same color and numerous stones occur in many colors. Color is produced by light; light is an electromagnetic vibration at certain wavelengths. The human eye can only perceive wavelengths between 750 and 380 nm (see page 35). This visible field is divided into several sectors of certain wavelengths, each of a particular color (spectral colors: red, orange, yellow, green, blue, violet).

Crystal with scepter amethyst; Mexico (somewhat reduced).

The mixture of all these colors produces white light. If, however, a certain wavelength (i.e., also the corresponding color) is absorbed out of the entire spectrum, the remaining mixture produces a certain color, but not white. If all wavelengths pass through the stone, it appears colorless. If all light is absorbed, the stone appears black. If all wavelengths are absorbed to the same degree, the stone is dull white or gray.

In the case of gemstones, the metals and their combinations, especially chrome, iron, cobalt, copper, manganese, nickel, and vanadium, absorb certain wavelengths of light and so cause coloration. In the case of zircon and smoky quartz, no impurity, or foreign substance, is responsible for the color, but rather a deformation of the internal crystal structure (lattice) results in the selective absorption of light, giving a change in the original color.

27

The distance the light ray travels through the stone can also influence absorption and thus color. The cutter must therefore use this fact to his advantage. Light-colored stones are made thicker and/or are given such an arrangement of facets that the absorption path lengthens, giving a deeper color. Materials with colors that are too dark are cut thinly. The dark red almandite garnet, for example, is therefore often hollowed out on the underside.

Artificial light has an influence on the color of gemstones, as it is usually differently composed than daylight. There are gemstones whose color is influenced unfavorably by artificial incandescent light (e.g., sapphire), and those which have an especially radiating effect in such light (e.g., ruby and emerald). The most obvious change in color occurs in alexandrite, which is green in daylight and red in artificial incandescent light.

Although color is of great importance in gems, with the exception of diamonds, no practical method of objective color determination is widely accepted. Color comparison charts are poor substitutes because there is too much room for subjective consideration. The measuring methods used in science for color determination are too complicated for the trade.

Color of Streak

The color appearance of gems, even in the same species, can vary greatly. For instance, beryl can have all the colors of the spectrum, but can also be colorless. This colorlessness is, in fact, the true color. It is called *inherent color*. All other colors are produced by impurities.

The inherent color, as it is constant, can help to identify a rough stone. This color can be seen by streaking the mineral on a rough porcelain plate, called the *streak plate*. The finely ground powder has the same effect as thin transparent platelets, from which the color-producing impurities have been abstracted. Steely hematite, for instance, has a streak color (called *streak*) which is red. Brass-colored pyrite's streak is black, and blue sodalite's is white. In the case of very hard gemstones, it is advisable first to remove a little powder with a steel file, and then rub it on the streak plate. This method of determination is of special interest to collectors. Because of the danger of damage, a cut gem should never be tested for streak. Refer to the table of streaks on page 30.

Changes in Color

The color of some gems is altered by time. Amethyst, rose quartz, and kunzite can become paler when exposed to direct sunlight. Generally color changes effected by natural causes are not common. Much more frequently man uses scientific methods to enhance the color of certain gemstones.

Best known is the heat treatment "firing" of amethyst. At several hundred degrees, the original violet stone becomes light yellow, red-brown, green, or milky white. Most citrines on sale and all prasiolites are amethysts whose color has been changed this way. Less attractive colors can be changed to more desirable hues by heating. Greenish aquamarines are heated to a sea-blue color; tourmalines which are too dark can be lightened; and blue tourmalines can be turned green. Colorless and aquamarine-colored zircons are produced by heating red-brown stones. Most rubies and blue sapphires are also heat treated to improve their color.

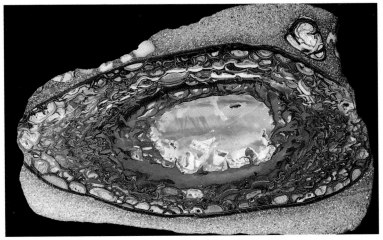

Opal in matrix, so-called boulder opal; Queensland, Australia (about natural size).

Colors can also be improved by radium and X-ray treatment and, more recently, by bombardment with elementary particles. The resulting colors are usually so close to nature that they cannot be detected by the eye; complicated tests are required to unmask them. Some of these resulting colors are not permanent; the stones can become pale, change color, or become spotty. In case of porous gems, such as lapis lazuli, turquoise, pearls, and agate, colors are improved by the addition of pigments, oils, resins, plastics, and wax. Such stone treatment is a very ancient practice.

Requirements for the Trade   In the trade, all artificially caused color changes of gemstones must be marked as such according to the guidelines of the CIBJO, with the exception of the following: those "which have received through heat treatment a permanent and irreversible color change" from the list "amber, beryl (aquamarine, morganite), corundum (sapphire, ruby), quartz (citrine, prasiolite, amethyst), topaz (rose topaz), tourmaline (all colors), zoisite (blue tanzanite)"; also in the case of those "which through heat treatment and through the influence of acid and corrosive mordant have received a permanent and irreversible change in color, such as banded agate, carnelian, onyx, green agate, and blue agate." These specific exceptions do not require the indication of artificial coloration in the trade.

Section 23.22 of the US FTC's *Guides for the Jewelry, Precious Metals, and Pewter Industries* states: "It is unfair or deceptive to fail to disclose that a gemstone has been treated in any manner that is not permanent or that creates special care requirements, and to fail to disclose that the treatment is not permanent, if such is the case. The following are examples of treatments that should be disclosed because they usually are not permanent or create special care requirements: coating, impregnation, irradiating, heating, use of nuclear bombardment, application of colored or colorless oil or epoxy-like resins, wax, plastic, or glass, surface diffusion, or dyeing. This disclosure may be made at the point of sale, except that disclosure should be made in any solicitation where the product can be purchased without viewing (e.g., direct mail catalogs, on-line services), and in the case of televised shopping programs, on the air. If special care requirements for a gemstone arise because the gemstone has been treated, it is recommended that the seller disclose the special care requirements to the purchaser."

Selected Gemstones
Ordered by Streak

White, colorless, gray
Actinolite
Agate
Alabaster
Alexandrite
Almandite
Amazonite
Amber
Amblygonite
Amethyst
Amethyst quartz
Anatase
Andalusite
Andradite
Anhydrite
Apatite
Apophyllite
Aquamarine
Aragonite
Augelite
Aventurine
Aventurine feldspar
Axinite
Barite
Barytocalcite
Benitoite
Beryllonite
Brazilianite
Calcite
Cancrinite
Carnelian
Cassiterite
Cerussite
Chalcedony
Charoite
Chrysoberyl
Chrysoprase
Citrine
Celestite
Colemanite
Coral
Cubic zirconia
Danburite
Datolite
Demantoid
Diamond
Diopside
Dolomite
Dumortierite
Emerald
Enstatite
Epidote
Euclase
Fluorite
Gahnite

Gaylussite
GGG
Glass
Grossularite
Hambergite
Haüynite
Hemimorphite
Hessonite
Hiddenite
Howlite
Hypersthene
Idocrase
Iolite
Ivory
Jadeite
Jasper
Kornerupine
Kunzite
Kurnakovite
Kyanite
Labradorite
Lazulite
Leucite
Linobate
Magnesite
Meerschaum
Milarite
Mimetite
Moldavite
Monazite
Moonstone
Montebrasite
Moss agate
Natrolite
Nephrite
Obsidian
Opal
Orthoclase
Parisite
Periclase
Peridot
Peristerite
Pearl
Petalite
Phenakite
Phosgenite
Prasiolite
Precious beryl
Prehnite
Pyrope
Quartz
Rhodochrosite
Rhodonite
Rock crystal
Rose quartz
Ruby

Sanidine
Sapphire
Scapolite
Scheelite
Serpentine
Siderite
Silver
Sinhalite
Smithsonite
Smoky quartz
Sodalite
Spessartite
Sphene
Spinel
Spodumene
Staurolite
Strontium titanate
Tanzanite
Topaz
Tourmaline
Tremolite
Turquoise
Tugtupite
Ulexite
Uvarovite
Variscite
Willemite
Witherite
Wulfenite
YAG
Zircon
Zoisite

Red, pink, orange
Cinnabar
Crocoite
Cuprite
Friedelite
Greenockite
Hematite
Manganotantalite
Piemontite
Proustite
Purpurite
Pyrargyrite
Pyroxmangite
Realgar

Yellow, orange, brown
Chromite
Descloizite

Durangite
Fergusonite
Gold
Hubnerite
Jet
Neptunite
Rutile
Sulfur
Sphalerite
Stibiotantalite
Thorianite
Tiger's-eye
Vanadinite
Wurtzite
Zincite

Green, yellow-green, blue-green
Bayldonite
Chrysocolla
Dioptase
Gadolinite
Gaspeite
Hornblende
Malachite
Marcasite

Blue, blue-green, blue-red
Azurite
Boleite
Ceruleite
Euzenite
Lapis lazuli
Linarite
Shattuckite
Vivianite

Black, gray
Anthophyllite
Aschynite
Bixbyite
Chalcopyrite
Davidite
Ilmenite
Ilvaite
Magnetite
Melonite
Pyrite
Pyrolusite
Tantalite
Wolframite

Refraction of Light

Most of us, when children, noticed that when a stick was partially immersed in water at a slant, it appeared to "break" at water level. The lower part of the stick appeared to be at a different angle from the upper part. What we observed was caused by the refraction of light. It always occurs when a ray of light leaves one medium (for instance, air) and enters obliquely into another (for instance, water and/or a gem crystal) at the interface between the two media.

The amount of refraction in the crystals is constant for each specific gemstone. It can therefore be used in the identification of the type of stone. The amount of the refraction is called the refractive index and is defined as the proportional relation between the speed of light in air to that in the stone. A decrease in the velocity of light in the stone causes a deviation of the light rays.

Example: Speed of light in air (V_1)   299 700 km/sec
Speed of light in diamond (V_2)   124 120 km/sec

$$\text{Refractive index} = \frac{V_1 \text{ (air)}}{V_2 \text{ (diamond)}} = \frac{299\ 700}{124\ 120} = 2.415$$

This means that the speed of light in air is 2.4 times faster than the speed of light in diamond. The refractive indices of gems are betwen 1.4 and 3.2 . They vary some-what with color and occurrence. Doubly refractive gems (see the descriptions on page 34) have two refractive indices. (Also refer to the refractive indices chart on pages 32 and 33.)

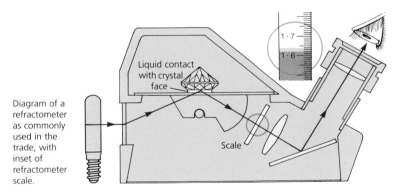

Diagram of a refractometer as commonly used in the trade, with inset of refractometer scale.

Liquid contact with crystal face

Scale

Refractometer

The light refraction is measured in practice with a refractometer. The values can be read directly from a scale. However, testing is only possible up to a value of 1.81 on a common instrument, and only stones with a flat face or facet are suitable. The determination of the refractive indices of other stones and of values over 1.81 requires the use of special devices.

The expert can find approximate values of cabachons using his own experience and knowledge.

Refractive Indices and Double Refraction for Selected Gemstones

	Refractive Index	Double Refraction		Refractive Index	Double Refraction
Hematite	2.940–3.220	0.287	Hessonite	1.730–1.757	none
Cinnabar	2.905–3.256	0.351	Epidote	1.729–1.768	0.015–0.049
Prousite	2.881–3.084	0.203	Azurite	1.720–1.848	0.108–0.110
Pyrargyrite	2.88–3.08	0.200	Pyrope	1.720–1.756	none
Cuprite	2.849	none	Hodgkinsonite	1.719–1.748	0.022–0.026
Rutile	2.616–2.903	0.287	Taaffeite	1.719–1.730	0.004–0.009
Brookite	2.583–2.700	0.117	Rhodonite	1.716–1.752	0.010–0.014
Anatase	2.488–2.564	0.046–0.067	Gahnospinel	1.715–1.754	none
Diamond	2.417–2.419	none	Spinel	1.712–1.762	none
Strontium titarate	2.409	none	Kyanite	1.710–1.734	0.015–0.033
			Adamite	1.708–1.760	0.048–0.050
Stibiotantalite	2.370–2.450	0.080	Diaspore	1.702–1.750	0.048
Sphalerite	2.368–2.371	none	Serendibite	1.701–1.743	0.005
Crocoite	2.29–2.66	0.270	Sapphirine	1.701–1.734	0.004–0.007
Wulfenite	2.280–2.400	0.120	Aegirine–augite	1.700–1.800	0.030–0.050
Tantalite	2.26–2.43	0.160	Idocrase	1.700–1.723	0.002–0.012
Linobate	2.202–2.273	0.071	Tanzanite	1.691–1.700	0.009
Mangano–tantalite	2.19–2.34	0.150	Neptunite	1.690–1.736	0.029–0.045
			Willemite	1.690–1.723	0.028–0.033
Mimetite	2.120–2.135	0.015	Rhodizite	1.690	none
Phosgenite	2.114–2.145	0.028	Triphylite	1.689–1.702	0.006–0.008
Cubic zirconia	2.088–2.176	none	Lithiophilite	1.68–1.70	0.01
Senarmontite	2.087	none	Dumortierite	1.678–1.689	0.015–0.037
GGG	2.03	none	Legrandite	1.675–1.740	0.060
Zincite	2.013–2.029	0.016	Hypersthene	1.673–1.731	0.010–0.016
Cassiterite	1.997–2.098	0.096–0.098	Parisite	1.671–1.772	0.081–0.101
Simpsonite	1.976–2.034	0.058	Clinozoisite	1.670–1.734	0.005–0.015
Sulfur	1.958–2.245	0.291	Sinhalite	1.665–1.712	0.036–0.042
Bayldonite	1.95–1.99	0.040	Lawsonite	1.665–1.686	0.019–0.021
Scheelite	1.918–1.937	0.010–0.018	Diopside	1.664–1.730	0.024–0.031
Andradite	1.88–1.94	none	Bustamite	1.662–1.707	0.014–0.015
Anglesite	1.878–1.895	0.017	Kornerupine	1.660–1.699	0.012–0.017
Uvarovite	1.87	none	Hiddenite	1.660–1.681	0.014–0.016
Purpurite	1.85–1.92	0.007	Kunzite	1.660–1.681	0.014–0.016
Sphene	1.843–2.110	0.100–0.192	Boracite	1.658–1.673	0.010–0.011
YAG	1.833	none	Axinite	1.656–1.704	0.010–0.012
Zircon	1.810–2.024	0.002–0.059	Malachite	1.655–1.909	0.254
Cerussite	1.804–2.079	0.274	Jadeite	1.652–1.688	0.020
Gahnite	1.791–1.818	none	Peridot	1.650–1.703	0.036–0.038
Spessartite	1.790–1,820	none	Ludlamite	1.650–1.697	0.038–0.044
Painite	1.787–1.816	0.029	Enstatite	1.650–1.680	0.009–0.012
Almandite	1.770–1.820	none	Euclase	1.650–1.677	0.019–0.025
Gadolinite	1.77–1.82	0.01–0.04	Phenakite	1.650–1.670	0.016
Ruby	1.762–1.778	0.008	Dioptase	1.644–1.709	0.051–0.053
Sapphire	1.762–1.778	0.008	Jet	1.640–1.680	none
Benitoite	1.757–1.804	0.047	Eosphorite	1.638–1.671	0.028–0.035
Shattuckite	1.752–1.815	0.063	Spurrite	1.637–1.681	0.039–0.040
Chrysoberyl	1.746–1.763	0.007–0.011	Jeremejevite	1.637–1.653	0.007–0.013
Periclase	1.74	none	Barite	1.636–1.648	0.012
Scorodite	1.738–1.768	0.027–0.030	Durangite	1.634–1.685	0.051
Staurolite	1.736–1.762	0.010–0.015	Siderite	1.633–1.875	0.242
Grossularite	1.734–1.759	none	Danburite	1.630–1.636	0.006–0.008
Pyroxmangite	1.734–1.756	0.017–0.019	Clinohumite	1.629–1.674	0.028–0.041
Chambersite	1.732–1.745	0.010	Apatite	1.628–1.649	0.002–0.006

	Refractive Index	Double Refraction		Refractive Index	Double Refraction
Andalusite	1.627–1.649	0.007–0.013	Petrified wood	1.54	none
Friedelite	1.625–1.664	0.030			
Smithsonite	1.621–1.849	0.228	Jasper	1.54	none
Datolite	1.621–1.675	0.040–0.050	Amber	1.539–1.545	none
Celestite	1.619–1.635	0.010–0.012	Ivory	1.535–1.570	none
Tourmaline	1.614–1.666	0.014–0.032	Apophyllite	1.535–1.537	0.002
Actinolite	1.614–1.653	0.020–0.025	Tiger's-eye	1.534–1.540	none
Hemimorphite	1.614–1.636	0.022	Pearls	1.53–1.69	0.16
Lazulite	1.612–1.646	0.031–0.036	Aragonite	1.530–1.685	0.155
Prehnite	1.611–1.669	0.021–0.039	Agate	1.530–1.540	0.004
Turquoise	1.610–1.650	0.040	Chalcedony	1.530–1.540	0.004
Topaz	1.609–1.643	0.008–0.016	Chrysoprase	1.530–1.540	0.004
Sugilite	1.607–1.611	0.001–0.004	Moss agate	1.530–1.540	0.004
Sogdianite	1.606–1.608	0.002	Meerschaum	1.53	none
Brazilianite	1.602–1.623	0.019–0.021	Witherite	1.529–1.677	0.148
Rhodochrosite	1.600–1.820	0.208–0.220	Milarite	1.529–1.551	0.003
Odontolite	1.60–1.64	0.010	Nepheline	1.526–1.546	0.0004
Nephrite	1.600–1.627	0.027	Aventurine feldspar	1.525–1.548	0.010
Pectolite	1.595–1.645	0.038			
Montebrasite	1.594–1.633	0.22	Amazonite	1.522–1.530	0.008
Phosphophyllite	1.594–1.621	0.021–0.033	Ammonite	1.52–1.68	0.155
Melinophane	1.593–1.612	0.019	Strontianite	1.52–1.67	0.150
Eudialyte	1.591–1.633	0.003–0.010	Gypsum	1.520–1.529	0.009
Chondrodite	1.59–1.64	0.035	Sanidine	1.518–1.530	0.008
Catapleiite	1.590–1.629	0.039	Moonstone	1.518–1.526	0.008
Wardite	1.590–1.599	0.009	Pollucite	1.517–1.525	none
Ekanite	1.590–1.596	0.001	Stichtite	1.516–1.544	0.026
Herderite	1.587–1.627	0.023–0.032	Thomsonite	1.515–1.542	0.006–0.025
Colemanite	1.586–1.615	0.028–0.030	Magnesite	1.509–1.717	0.022
Howlite	1.586–1.605	0.019	Scolecite	1.509–1.525	0.007–0.012
Zektzerite	1.582–1.585	0.003	Leucite	1.504–1.509	none
Amblygonite	1.578–1.646	0.024–0.030	Mesolite	1.504–1.508	0.001
Anhydrite	1.570–1.614	0.044	Dolomite	1.502–1.698	0.185
Augelite	1.570–1.590	0.014–0.020	Petalite	1.502–1.519	0.012–0.017
Emerald	1.565–1.602	0.006	Lapis lazuli	1.50	none
Aquamarine	1.564–1.596	0.004–0.005	Haüynite	1.496–1.510	none
Variscite	1.563–1.594	0.031	Tugtupite	1.496–1.502	0.006
Precious beryl	1.562–1.602	0.004–0.010	Cancrinite	1.495–1.528	0.024–0.029
Tremolite	1.560–1.643	0.017–0.027	Ulexite	1.491–1.520	0.029
Vivianite	1.560–1.640	0.050–0.075	Moldavite	1.49–1.51	none
Serpentine	1.560–1.571	0.008–0.014	Yugawaralite	1.490–1.509	0.011–0.014
Labradorite	1.559–1.570	0.008–0.010	Whewellite	1.489–1.651	0.159–0.163
Hambergite	1.553–1.628	0.072	Kurnakovite	1.488–1.525	0.036
Beryllonite	1.552–1.561	0.009	Inderite	1.488–1.515	0.017–0.027
Charoite	1.550–1.561	0.004–0.009	Calcite	1.486–1.658	0.172
Amethyst	1.544–1.553	0.009	Coral	1.486–1.658	0.172
Aventurine	1.544–1.553	0.009	Natrolite	1.480–1.493	0.013
Rock Crystal	1.544–1.553	0.009	Sodalite	1.48	none
Citrine	1.544–1.553	0.009	Analcite	1.479–1.489	none
Prasiolite	1.544–1.553	0.009	Thaumasite	1.464–1.507	0.036
Smoky quartz	1.544–1.553	0.009	Creedite	1.461–1.485	0.024
Rose quartz	1.544–1.553	0.009	Chrysocolla	1.460–1.570	0.023–0.040
Andesine	1.543–1.551	0.008	Obsidian	1.45–1.55	none
Iolite	1.542–1.578	0.008–0.012	Gaylussite	1.443–1.523	0.080
Oligoclase	1.542–1.549	0.007	Glass	1.44–1.90	none
Talc	1.54–1.59	0.050	Fluorite	1.434	none
Scapolite	1.540–1.579	0.006–0.037	Sellaite	1.378–1.390	0.010–0.012
Amethyst quartz	1.54–1.55	0.009	Opal	1.37–1.52	none

Gemstones in an immersion liquid.

1. White contour and dark facet edges: gem has lower refractive index.

2. Black contour and white facet edges: gem has higher refractive index.

3. Widened contour: refractive indices vary considerably.

4. Indistinct outline (tending to disappear): liquid and gem have same refractive index.

(According to Dragstedt et al.)

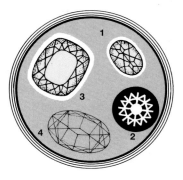

Immersion Method   Without access to a costly apparatus, a rough measurement of the refractive index is fairly easy with the immersion method. The gem is viewed after immersing in a liquid with a known refractive index. On the basis of brightness, sharpness, and width of the contours/outlines as well as the facet edges, one can learn approximate values about the refractive index.

Double Refraction

In all gemstones, except opals, glasses, and those belonging to the cubic system, the ray of light is refracted when entering the crystal and at the same time divided into two rays. This phenomenon is called double refraction. It can be most clearly observed in the case of calcite. It is also easily seen in zircon, sphene, tourmaline, and peridot. When looking from above, one can see the doubling of the edges of the lower facets in transparent cut stones. Often, magnification is needed to see the effect. It is up to the lapidary to work the stone in such a way that the double refraction does not appear disturbing.

The double refraction can be useful in identifying gemstones. It is expressed as the difference between the highest and lowest refractive index. The expert also differentiates between positive and negative "optical character." (Refer to the table on pages 32 and 33.)

Calcite shows double refraction especially clearly (left), and diagram of double refraction (right).

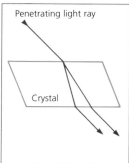

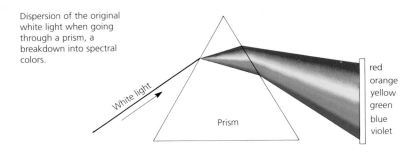

Dispersion of the original white light when going through a prism, a breakdown into spectral colors.

White light

Prism

red
orange
yellow
green
blue
violet

Dispersion

In colorless and cut gemstones, one can occasionally observe flashes of color, which come about through dispersion of the white light into the spectral colors. That is to say, the white light is not only refracted when penetrating a crystal, but it is also dispersed into its spectral colors, because each wavelength is refracted by a different amount. Violet is refracted more strongly than red. The dispersion is different from one gemstone to another. A distinct dispersion usually occurs only in colorless or weakly tinted stones. Facets can enhance the dispersion. Colorful gemstones tend to mask the dispersion effect.

Color dispersion is especially high in diamonds where it produces the so-called fire. Natural as well as synthetic gemstones with high dispersion (for instance, strontium titanate, synthetic rutile, sphalerite, sphene, and zircon) are used as substitutes for diamond; sometimes, though, they are supposititious (see page 75).

The measurement of the dispersion can be done with a refractometer and special devices. (However, this is not part of routine gem testing.) The dispersion of a stone is expressed in figures as the difference between the red and violet refractive indices. As the color usually comprises a wide spectrum, it is common to use certain lines (Fraunhofer lines) of the spectrum when doing the measurements. In the gemological literature, mainly the Fraunhofer lines B and G (BG dispersion) are used as a basis for the dispersion values, but occasionally the lines C and F (CF dispersion) are used.

Tables of dispersion values can be consulted for the determination of the gemstone. In the table on page 36, the BG dispersion values are contrasted with those of the CF dispersion. In the description of the individual gemstones, the CF dispersion values are placed in brackets following the BG values.

It is important to remember that only transparent stones can show any dispersion at all. Additionally, in gemstones with dispersion lower than zircon's (0.039), the effect may not be visible except in large and colorless or very lightly tinted specimens.

Spectrum of the Fraunhofer lines. Dispersion refers either to the BG or the CF area.

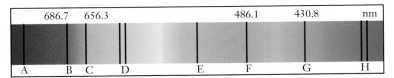

| | 686.7 | 656.3 | | 486.1 | 430.8 | nm |
| A | B | C | D | E | F | G | H |

Dispersion
of Selected Gemstones

	B-G	C-F		B-G	C-F
Synthetic Rutile	0.280	0.120–0.180	Herderite	0.017	0.008–0.009
Anatase	0.213–0.259		Hiddenite	0.017	0.010
Wulfenite	0.203	0.133	Kunzite	0.017	0.010
Strontium titanate	0.19	0.109	Scapolite	0.017	
			Spodumene	0.017	0.010
Sphalerite	0.156	0.088	Tourmaline	0.017	0.009–0.011
Linobate	0.120	0.075	Andalusite	0.016	0.009
Cassiterite	0.071	0.035	Barite	0.016	0.009
Cubic zirconia	0.065	0.035	Datolite	0.016	
Andradite	0.057		Euclase	0.016	0.009
Demantoid	0.057		Chrysoberyl	0.015	0.011
Cerussite	0.055	0.033–0.050	Hambergite	0.015	0.009–0.010
Sphene	0.051	0.019–0.038	Phenakite	0.015	0.009
Benitoite	0.046	0.026	Rhodochrosite	0.015	0.010–0.020
Anglesite	0.044	0.025	Sillimanite	0.015	
Diamond	0.044	0.025	Smithsonite	0.014–0.031	0.008–0.017
Zircon	0.039	0.022	Amblygonite	0.014–0.015	0.008
GGG	0.038	0.022	Aquamarine	0.014	0.009–0.013
Scheelite	0.038	0.026	Brazilianite	0.014	0.008
Dioptase	0.036	0.021	Precious beryl	0.014	0.009–0.013
Whewellite	0.034		Emerald	0.014	0.009–0.013
Gypsum	0.033	0.008	Topaz	0.014	0.008
Epidote	0.030	0.012–0.027	Amethyst	0.013	0.008
Tanzanite	0.030	0.011	Amethyst quartz	0.013	0.008
YAG	0.028	0.015	Apatite	0.013	0.010
Almandite	0.027	0.013–0.016	Aventurine	0.013	0.008
Hessonite	0.027		Rock crystal	0.013	0.008
Spessartite	0.027	0.015	Citrine	0.013	0.008
Willemite	0.027		Prasiolite	0.013	0.008
Boracite	0.024	0.012	Quartz	0.013	0.008
Staurolite	0.023	0.012–0.013	Smoky quartz	0.013	0.008
Pyrope	0.022	0.013–0.016	Feldspar	0.012	0.008
Grossularite	0.020	0.012	Pollucite	0.012	0.007
Hemimorphite	0.020	0.013	Beryllonite	0.010	0.007
Kyanite	0.020	0.011	Cancrinite	0.010	0.008–0.009
Peridot	0.020	0.012–0.013	Leucite	0.010	0.008
Spinel	0.020	0.011	Strontianite	0.008–0.028	
Idocrase	0.019–0.025	0.014	Calcite	0.008–0.017	0.013–0.014
Clinozoisite	0.019	0.011–0.014	Fluorite	0.007	0.004
Labradorite	0.019	0.010	Gahnite		0.019–0.021
Axinite	0.018–0.020	0.011	Uvarovite		0.014–0.021
Ekanite	0.018	0.012	Dolomite		0.013
Kornerupine	0.018	0.010	Enstatite		0.010
Corundum	0.018	0.011	Phosphophyllite		0.010–0.011
Ruby	0.018	0.011	Actinolite		0.009
Sapphire	0.018	0.011	Jeremejevite		0.009
Sinhalite	0.018	0.010	Apophyllite		0.008
Sodalite	0.018	0.009	Celestite		0.008
Diopside	0.017–0.020	0.012	Haüynite		0.008
Iolite	0.017	0.009	Natrolite		0.008
Danburite	0.017	0.009	Aragonite		0.007–0.012
			Tremolite		0.006–0.007

Absorption Spectra

The absorption spectrum of a gem can sometimes be one of the most important aids in identifying it. The absorption spectrum of a stone consists of the bands that appear in the spectral colors of light as they emerged from the gemstone (see page 40). As we can see, certain wavelengths (color bands) of the light are absorbed (see page 27) and the color of the gem is formed from the mixture of the remaining parts of the original white light. The human eye cannot recognize all minute color differences. Red tourmaline, for instance, can appear like red garnet, or even red-colored glass can appear deceivingly like the desirable red ruby. However, the absorption spectrum unmasks without any doubt the stone or glass used to imitate the ruby. Many gems have a very characteristic, even unique, absorption spectrum which is revealed (seen through a spectroscope) in black vertical lines or broad bands.

The great advantage of this testing method is the ease with which one can differentiate between gems of the same density and similar refractive index. One can also use this method for testing rough stones, cabochons, and even set stones. An important area where absorption spectroscopy is applied is in differentiating between natural stones, synthetic stones, and imitations.

Best results are obtained from strongly colored, transparent gemstones. The observation of the absorption spectrum of opaque stones is only possible when a very thin slice of the stone is prepared which can transmit light. Otherwise translucent edge must be presented, or light must be reflected from the surface.

The testing instrument is the spectroscope, with the help of which one can determine the wavelength of the absorbed light. The wavelength is measured in nanometers, symbol nm ($1 nm = 10^{-9}$ m = 1 millionth millimeter). The former measurement Ångstrom, symbol Å ($1 Å = 10^{-10}$ m = 0.1 nm) is still widely used in the gemological literature. Because the absorption lines or bands are not always of equal strength, it is usual to note such measured differences. Strong absorption lines are in this book underlined, for instance, <u>653</u>; medium strong absorption lines are in normal lettering, for instance 594; weak lines are bracketed, for instance (432). (See the table of the absorption spectra of selected gemstones on pages 38 and 39.)

Hand spectroscope with wavelength scale in the small tube.

Absorption Spectra of Selected Gemstones

All figures are in nanometers (nm).
Strong absorption lines are underlined; weak ones are in parentheses.

Actinolite: 503, 431
Agate, artificially dyed green: <u>700</u>, (665), (634)
Alexandrite, green direction: <u>680</u>, 678, 665, <u>655</u>, 649, 645, <u>640</u>–<u>555</u>
Alexandrite, red direction: 680, <u>678</u>, 655, 645, 605–540, (472)
Almandite: 617, <u>576</u>, <u>526</u>, <u>505</u>, 476, 462, 438, 428, 404, 393
Amethyst: (550–520)
Andalusite: 553, <u>550</u>, 547, (525), (518), (495), <u>455</u>, 447, <u>436</u>
Andradite: <u>701</u>, 693, <u>640</u>, <u>622</u>, <u>443</u>
Apatite, blue: <u>512</u>, <u>507</u>, <u>491</u>, 464
Apatite, yellow-green: 597, <u>585</u>, <u>577</u>, 533, 529, 527, 525, 521, 514, 469
Aquamarine: <u>537</u>, 456, 427
   Maxixe: 695, <u>655</u>, 628, 615, 581, 550
Aventurine, green: 682, 649
Axinite: 532, <u>512</u>, <u>492</u>, <u>466</u>, 440, <u>415</u>
Azurite: 500
Calcite: <u>582</u>
Chalcedony, artificially dyed blue: 690–660, 627
Chalcedony, artifically dyed green: 705, 670, 645
Chrysoberyl: 504, (495), 485, <u>445</u>
Chrysoprase, artificially dyed with nickel: 632, 444
Chrysoprase, natural: 444
Cubic zirconia, orange: <u>640</u>, <u>630</u>, (540), (536), (533), (530), <u>520</u>, <u>517</u>, <u>515</u>, <u>512</u>, <u>510</u>, (503), <u>482</u>, <u>480</u>, <u>477</u>, <u>475</u>, (449), (447), (446)
Danburite: 590, 586, <u>585</u>, 584, 583, 582, 580, 578, 573, 571, 568, 566, 564
Demantoid: 701, 693, <u>640</u>, <u>622</u>, <u>443</u>
Diamond, artificially colored brown: (741), 594, <u>504</u>, <u>498</u>, 478, 465, 451, 435, 423, 415
Diamond, artificially colored green: <u>741</u>, <u>504</u>, 498, 465, 451, 435, 423, 415
Diamond, artificially colored yellow: <u>594</u>, 504, 498, (478), (415)
Diamond, natural brown-green: (537), <u>504</u>, (498)
Diamond, natural colorless to yellow (Cape): <u>478</u>, 465, 451, 435, 423, <u>415</u>, 401, 390
Diamond, natural yellow-brown: 576, 569, 564, 558, 550, 548, 523, 493, 480, 460
Diopside: (505), (493), (446)
   Chrome Diopside: (690), (670), (655), (635), <u>508</u>, <u>505</u>, 490
Dioptase: <u>550</u>, <u>465</u>
Ekanite: 665, (637)
Emerald, natural: <u>683</u>, <u>681</u>, 662, 646, <u>637</u>, (606), (594), <u>630</u>–<u>580</u>, 477, 472
Emerald, synthetic: <u>683</u>, <u>680</u>, 662, 646, <u>637</u>, 630–580, 606, 594, 477, 472, <u>430</u>
Enstatite: <u>547</u>, 509, <u>505</u>, 502, 483, 459, 449
   Chrome–Enstatite: 688, 669, 506
Epidote: 475, <u>455</u>, 435
Eosphorite, brownish pink: 490, <u>410</u>
Euclase: <u>706</u>, <u>704</u>, <u>650</u>, <u>639</u>, 468, 455
Fluorite, green: 634, 610, 582, 445, 427
Fluorite, yellow: 545, 515, 490, 470, 452
Friedelite: 556, (456)
Gahnite: <u>632</u>, 592, 577, 552, 508, <u>480</u>, <u>459</u>, 443, 433
Grossularite: 697, <u>630</u>, 605, 505
Hematite: (700), (640), (595), (570), (480), (450), (425), (400)
Hessonite: 547, 490, 454, <u>435</u>
Hiddenite: <u>690</u>, <u>686</u>, 669, 646, <u>620</u>, <u>437</u>, 433
Hypersthene: 551, <u>547</u>, <u>505</u>, 482, 448
Idocrase, brown: 591, 588, <u>584</u>, 582, 577, 574

38

Idocrase, green: (528), <u>461</u>
Idocrase, yellow–green: <u>465</u>
Iolite: 645, 593, 585, 535, <u>492</u>, <u>456</u>, 436, 426
Jadeite, artificially dyed green: 665, 655, 645
Jadeite, natural green: <u>691</u>, 655, 630, (495), 450, <u>437</u>, 433
Kornerupine: 540, <u>503</u>, 463, <u>446</u>, 430
Kyanite: (706), (689), (671), (652), <u>446</u>, 433
Nephrite: (689), <u>509</u>, 490, 460
Obsidian, green: 680, 670, 660, 650, 635, 595, 555, 500
Opal, (fire–opal): 700–640, 590–400
Orthoclase: 448, 420
Peridot: <u>497</u>, <u>495</u>, <u>493</u>, <u>473</u>, <u>453</u>
Petalite: (454)
Precious Beryl, artificially dyed blue: 705–685, 645, 625, 605, (587)
Prehnite: 438
Pyrope: <u>687</u>, <u>685</u>, 671, 650, <u>620–520</u>, 505
Quartz, synthetic blue: 645, 585, 540, 500–490
Rhodochrosite: 551, 449, 415
Rhodonite: 548, 503, 455, (412), (408)
Ruby: <u>694</u>, <u>693</u>, 668, 659, 610–500, <u>476</u>, <u>468</u>, <u>465</u>
Sapphire, blue from Australia: <u>471</u>, <u>460</u>, <u>450</u>
Sapphire, blue from Sri Lanka: (450)
Sapphire, green: 471, 460–450
Sapphire, yellow: <u>471</u>, <u>460</u>, <u>450</u>
Scapolite, pink: 663, 652
Scheelite: <u>584</u>
Serpentine: 492, 464
Sillimanite: 462, 441, 410
Sinhalite: 526, 492, 475, <u>463</u>, 452
Sogdianite: (645–630), (493–488), <u>437</u>, <u>419</u>, <u>411</u>
Spessartite: 495, 485, 462, <u>432</u>, 424, <u>412</u>
Sphalerite: 690, 667, <u>651</u>
Sphene: <u>586</u>, <u>582</u>
Spinel, natural blue: <u>632</u>, <u>585</u>, <u>555</u>, 508, <u>478</u>, <u>458</u>, 443, 433
Spinel, natural red: <u>685</u>, <u>684</u>, <u>675</u>, <u>665</u>, 656, 650, 642, 632, <u>595–490</u>, 465, 455
Spinel, synthetic blue: 634, 580, 544, 485, 449
Spinel, synthetic cobalt blue: <u>635</u>, <u>580</u>, <u>540</u>, (478)
Spinel, synthetic green: 620, 580, 570, 550, 540
Spinel, synthetic yellow–green: 490, 445, 422
Stichtite: <u>665</u>, <u>630</u>
Sugilithe: 570, <u>419</u>, (411)
Taaffeite: 558, 553, 478
Tanzanite: <u>595</u>, 528, 455
Topaz, pink: <u>682</u>
Tourmaline, green: <u>497</u>, <u>461</u>, 415
Tourmaline, red: 555, 537, 525–461, <u>456</u>, <u>451</u>, 428
Tremolite: <u>684</u>, 650, 628
Turquoise: (460), 432, (422)
Variscite: 688, (650)
Verdite: 700, 699, 455
Willemite: 583, 540, 490, 442, 431, <u>421</u>
Zircon, high–zircon: 691, 689, 662, 660, <u>653</u>, 621, 615, 589, 562, 537, 516, 484, 460, 433
  low zircon: <u>653</u>, (520)

Absorption Spectra of Selected Gemstones

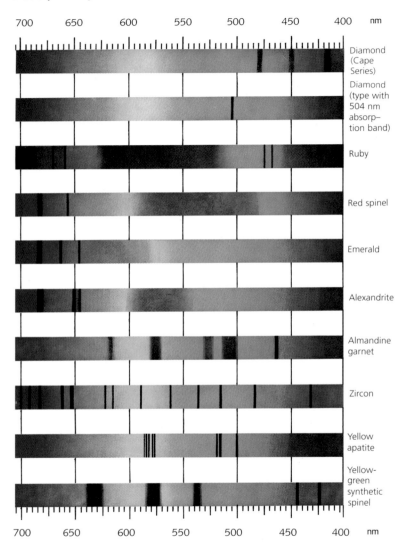

From *Gemmologists' Compendium* by R. Webster, NAG Press Ltd., London. Taken from *Edelsteinkundliches Handbuch* by Professor Dr. Chudoba and Dr. Gübelin, Stollfuss Verlag, Bonn.

Transparency

A factor in the evaluation of most gemstones is their transparency. Inclusions of foreign matter or fissures in the interior of the crystal affect the transparency or "clarity." The path of light through the crystal can also be impaired by strong absorption in the crystal. Grainy, stalky, or fibrous aggregates (such as chalcedony, lapis lazuli, and turquoise) are opaque because the rays of light are repeatedly refracted or reflected by the many tiny faces until finally they are completely reflected or absorbed. Where the light is only weakened by its passage through a stone, it is said to have translucency.

Luster and Brilliance

The luster of a gem is caused by external reflection, i.e., the reflecting of part of the incident light back from the surface. It is dependent on the refractive index and the nature of the surface, but not on the color. Generally, the higher the refraction, the higher the luster. The highest luster seen in transparent gemstones is adamantine (diamond-like); the most common is vitreous (glassy). Comparatively rare are the greasy, metallic, pearly, silky, and waxy lusters. Stones with no luster are described as dull.

Those light effects which are caused by total reflection are considered as brilliance or light return. In transparent faceted gemstones the lower facets act as a mirror and reflect the entering light more or less completely, thus creating the brilliance. The ideal complete internal reflection is found in the diamond cut, which thus reaches the highest brilliance.

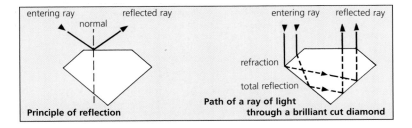

Principle of reflection

Path of a ray of light through a brilliant cut diamond

Pleochroism

Some gems appear to have different colors or depth of color when viewed in different directions. This is caused by the differing absorption of light rays in doubly refractive crystals. Where two main colors can be observed (only in the tetragonal, hexagonal, and trigonal crystal systems), one speaks of dichroism; where three colors can be seen (only in the orthorhombic, monoclinic, and triclinic crystal systems) of trichroism or pleochroism. The latter term is also a collective description used for both kinds of the multi-coloredness.

Amorphous gems and those of the cubic crystal system show no pleochroism. Opaque stones and most that are translucent also do not show it. The appearances of the pleochroism can be weak, definite, or strong. It must be taken into consideration when cutting in order to avoid poor colors, or shades that are too dark or too light. The instrument of observation for pleochroism is the dichroscope. (Refer to table of the pleochroism of selected gemstones on pages 42 and 43.)

Pleochroism
of Selected Gemstones

Actinolite:	yellow-green, light green, blue-green
Alexandrite:	green alexandrite viewed in daylight: distinct; pigeon-blood red-violet
	alexandrite in lamplight distinct: dark red, yellow-red, dark green
Anhydrite	violet crystals: colorless to pale yellow, pale violet to rose, violet
Amblygonite:	weak red, orange-yellow, emerald-green
Amethyst:	very weak; purple, gray-purple
Anatase:	distinct; yellow, orange
Andalusite:	strong; yellow, olive, red-brown to dark red
Anhydrite:	violet crystals: colorless to pale yellow, pale violet to rose, violet
Apatite:	yellow: weak; golden-yellow, green-yellow
	green: weak; yellow, green
	blue: very strong; blue, colorless
Aquamarine:	blue: distinct; nearly colorless to light blue, blue to sky-blue
	green/blue: distinct; yellow-green to colorless, blue-green
Axinite:	strong; olive-green, red-brown, yellow-brown
Azurite:	distinct; light blue, dark blue
Barite:	blue: weak
Benitoite:	very strong; colorless, blue
Brazilianite:	very weak
Cassiterite:	weak to strong; green-yellow, brown, red-brown
Charoite:	various: colorless, rose
Chrysoberyl:	very weak; red to yellow, yellow to pale green, green
Citrine:	natural: weak; yellow, light yellow
Danburite:	weak; weak pale yellow, light yellow
Diaspore:	bright: violet/blue, pale green, rose to dark red
Diopside:	weak; yellow-green, dark green
Dioptase:	weak; dark emerald-green, light emerald-green
Dumortierite:	strong; black, red-brown, brown
Emerald:	distinct; green, blue-green to yellow-green
Enstatite:	distinct; green, yellow-green
Euclase:	very weak; whitish-green, yellow-green, blue-green
Epidote:	green-brown epidote strong: green, brown, yellow;
	green epidote strong: almost colorless, light brown
	distinct: blue-green, emerald-green, yellow-green
Hiddenite:	distinct; blue-green, emerald-green, yellow-green
Hodgkinsonite:	various: multicolored, colorless
Hypersthene:	strong; hyacinth-red, straw-yellow, sky-blue
Idocrase:	green idocrase weak: yellow-green, yellow-brown;
	yellow idocrase weak: yellow, almost colorless;
	brown idocrase weak: yellow-brown, light-brown
Iolite:	blue iolite strong: yellow, dark blue-violet, pale blue;
	pale blue iolite strong; almost colorless, dark blue, pale blue
Kornerupine:	strong: green, yellow, reddish-brown
Kunzite:	distinct; amethyst color, pale red, colorless
Kyanite:	strong; pale blue to colorless, pale blue, dark blue
Lazulite:	strong; colorless, dark blue
Neptunite:	yellow, deep red
Orthoclase:	yellow: weak
Painite:	strong; ruby-red, brown-orange
Parisite:	pale
Peridot:	very weak; colorless to pale green, lively green, oily green
Phenakite:	distinct; colorless, orange-yellow

Prasiolite: very weak; light green, pale green
Precious beryl: gold-beryl weak: lemon-yellow, yellow;
                     green beryl: distinct: yellow-green, blue-green;
                     heliodor weak: golden-yellow, green-yellow;
                     morganite distinct: pale pink, blue-pink
Proustite   bright: red shadings
Purpurite: distinct; brown-gray, blood-red
Rhodonite: distinct: red-yellow, rose-red, red-yellow
Rose quartz: weak; pink, pale pink
Ruby: strong; yellow-red, deep carmine red
Rutile: various: red-brown, yellow, green
Sanidine: weak
Sapphire: orange: strong; yellow-brown to orange, nearly colorless
                     yellow: weak; yellow, light yellow
                     green: weak; green-yellow, greenish-yellow
                     blue: distinct; dark blue, greenish-blue
                     purple: distinct; purple, light red
Scapolite: pink: colorless, pink
                     yellow: distinct; colorless, yellow
Scheelite: variable
Sillimanite: strong; light green, dark green, blue
Sinhalite: distinct; green, light brown, dark brown
Smoky quartz: dark: distinct; brown, red-brown
Sphene: green sphene: colorless, green-yellow, reddish-yellow;
                     yellow sphene strong: colorless, green-yellow, reddish
Staurolite: strong; yellowish, yellowish-red, red
Tantalite: bright: brown, red-brown
Tanzanite: very strong; purple, blue, brown or yellow
Topaz: red: strong; dark red, yellow, rose-red
                     pink: distinct; colorless, pale pink, pink
                     yellow: distinct; lemon yellow, honey-yellow, straw-yellow
                     brown: distinct; yellow-brown, dull yellow-brown
                     green: distinct; pale green, light blue-green, greenish-white
                     blue: weak; light blue, pink, colorless
Tourmaline: red: distinct; dark red, light red
                     pink:distinct; light red, reddish yellow
                     yellow: distinct; dark yellow, light yellow
                     brown: distinct; dark brown, light brown
                     green: strong; dark blue, light blue
                     blue: strong; dark blue, light blue
                     purple: strong; purple, light purple
Tremolite: distinct
Tugtupite: bright: bluish-red, orange-red
Vivianite: strong: blue to indigo, pale yellow, green to blue-green, pale
                     yellowish-green
Willemite: variable
Zircon: red: weak; red, light yellow
                     red-brown: very weak; reddish-brown, yellowish-brown
                     red-brown: very weak; honey-yellow, brown-yellow
                     brown: very weak; red-brown, yellow-brown
                     brown-green: very weak; pink-yellow, lemon-yellow
                     green: very weak; green, brown-green
                     blue: distinct; blue, yellow-gray to colorless

Light and Color Effects

Many gems show striated light effects or color effects which do not relate to their body-color and are not caused by impurities or their chemical composition. These effects are caused by reflection, interference, and refraction.

Adularescence   Moonstone, being a variety of feldspar (see page 164), shows a blue-whitish opalescence (sometimes described as a "billowy" light) which glides over the surface when the stone is cut en cabochon. Interference phenomena of the layered structure are the cause of this effect, known as adularescence.

Asterism   This is the effect of light rays forming a star (Latin *aster*, star); the rays meet in one point and enclose definite angles (depending on the symmetry of the stone). It is usually created through reflection of light by thin fibrous or needle-like inclusions that lie in various directions. (See photograph on page 45.) Ruby (page 82) and sapphire (page 86) cabochons can show effective six-rayed stars. There are

Spectrolite, a variety of the labradorite with especially effective labradorization; Finland.

Quartz Cat's Eye. The finest rutile fibers cause the chatoyancy.

Light stars (asterism) in blue sapphires and in ruby.

also four-rayed stars and, rarely, twelve-rayed stars. If a piece of star rose quartz has been cut as a sphere, the rays move in circles over the whole surface; where included needles are partially destroyed, stunted stars, part circles, or light clusters are formed. Asterism also occurs in synthetic gems.

Aventurescence   Colorful play of glittering reflections of small, plate- or leaf-like inclusions. The inclusions are hematite or goethite in the case of aventurine feldspar (page 166); fuchsite or hematite in aventurine quartz (page 122); and copper scrapings in aventurine glass.

Chatoyancy (cat's-eye effect)   An effect which resembles the slit eye of a cat (French *chat* = cat, *oeil* = eye); this is caused by the reflection of light by parallel fibers, needles, or channels. This phenomenon is most effective when the stone is cut en cabochon in such a way that the base is parallel to the fibers. When the gem is rotated, the cat's-eye glides over the surface. The most precious cat's-eye is that of chrysoberyl (page 98). The effect can be found in many gemstones; especially well known are quartz cat's-eye, hawk's-eye, and tiger's-eye (page 124). If one talks simply of cat's-eye, one refers to a chrysoberyl cat's-eye. All other cat's-eye must have an additional designation.

Iridescence   The rainbow-like hues (Latin *iris* = rainbow) seen in some gems, caused by cracks or structural layers breaking up light into spectral colors. Fire agate (page 134) is a natural gemstone that shows this phenomenon. Commercially, it is created by artificially producing cracks in rock crystal.

45

Play of color in black precious opal; Australia.

Labradorescence   Iridescence in metallic hues, called schiller, found especially in labradorite (hence the name) and spectrolite (page 166). Blue and green effects are often found, but the whole spectrum can be observed. The cause of the schillers is most probably due to lattice distortions accompanying alternating microscopic exsolution lamellae of high- and low-calcium plagioclase.

Opalescence   Milky-blue or pearly appearance of common opal (page 152) caused by reflection of short wave, mainly blue, light. It should not be confused with play-of-color.

Orient   Iridescence in pearls. It is created through diffraction and interference of the light by the shingle-like layers of aragonite platelets near the pearl surface (see page 230).

Play-of-Color   Flashes of rainbow colors in opal (page 150) which change with the angle of observation. The electron-microscope shows the cause at a magnification of 20,000×: small spheres of the mineral crystobalite included in a silica gel cause the diffraction interference phenomena. The diameter of these spheres is one to two ten-thousandths of a millimeter.

Silk   Reflection of fibrous inclusions or canals causes a silk-like appearance. Especially desirable in faceted rubies and sapphires. Where the included needles are sufficiently numerous, the stone can display chatoyancy if cut en cabochon.

Luminescence

Luminescence (Latin *lumen*–light) is a collective term for the emission of visible light under the influence of certain rays, as well as by some physical or chemical reaction, but not including pure heat radiation.

The most important of these phenomena for the testing of gems is the luminescence under ultraviolet light, which is called fluorescence. The name fluorescence is derived from the mineral fluorite, which is the substance in which this light phenomenon was first observed. When the substance continues to give out light after irradiation has ceased, the effect is called phosphorescence (named after the well-known light property of phosphorus).

The causes of fluorescence are certain interference factors (impurities or flaws in the structure) in the crystal lattice. Many gemstones fluoresce to shortwave UV (254 nm). There are gemstones which exclusively react to shortwave, others only to longwave (366 nm), and again others which react to shortwave as well as longwave UV. Gemstones which contain iron do not show any fluorescence.

Fluorescence is not diagnostic because many specimans of a gemstone can fluoresce in completely different colors, while others of the same gemstone may not light up at all under UV. In the detection of synthetic gemstones, on the other hand, fluorescence can be determinative because synthesis under UV frequently react differently from natural gemstones. Also glued gemstones can sometimes be identified under UV when the glue fluoresces by itself or differently from the other parts. Diamonds (which fluoresce under X rays) can be separated from other stones (such as quartz or pyrope garnet). Occasionally, fluorescence can help establish a particular source locality, because sometimes typical phenotypes for a certain place are characteristic.

Luminescence caused by X rays can help to differentiate between natural and cultured pearls. The mother-of-pearl of saltwater pearl oysters does not luminesce while that of freshwater pearl mussels gives off a strong light, as the inserted nucleus of a cultured perl shows a luminescence which the natural ones do not have. (Refer to table of the fluorescing gemstones on page 48.) X-ray testing of pearls is done only by trade laboratories.

Fluorescing minerals in white light (left) and under UV rays (right). Each from the left; top, aragonite, calcite; center, fluorite, halite; bottom, willemite.

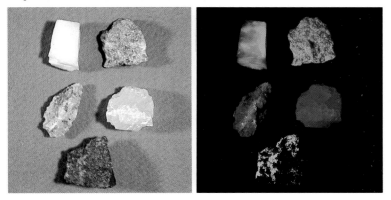

Fluorescence
of Selected Gemstones

Agate: differs within layers, partly strong;
  yellow, blue-white
Amazonite: weak; olive-green
Amber: bluish to yellow-green
  Burmite: blue
Amblygonite: very weak; green
Amethyst: weak; greenish
Ammonite: mustard-yellow
Analcime: creamy white
Andalusite: weak; green, yellow-green
Anglesite: weak; yellowish
Anhydrite: red
Apatite, yellow A.; violet to pink
Aragonite: weak; pinkish, yellow, yellow-
  brown, green, bluish (rare)
Aventurine feldspar: dark red-brown
Axinite: red, orange
Barite: white, blue-green, gray
Benitoite: strong; blue
Boracite: weak; greenish
Calcite: red, pink, orange, white, yellow-
  white
Cerussite: yellow, pink, green, bluish
Chalcedony: bluish-white
Chrysoberyl: green: weak; dark red,
  others: none
Colemanite: white, yellow-white
Coral: weak, violet
Danburite: sky-blue
Diamond: varies considerably
  colorless and yellow: usually blue
  brown and green: often green
  synthetic: strong; yellow
Diopside: violet, orange, yellow, green
Dolomite: pink, orange-red
Dumortierite: weak, blue, blue-white, violet
Emerald: usually none
Euclase: weak or none
Fluorite: usually strong; blue to violet
Gaylussite: weak; creamy white
Hambergite: usually none, once in a while
  orange
Hemimorphite: weak
Herderite: weak; green, violet
Hiddenite: very weak; red-yellow
Hodgkinsonite: weak, red
Howlite: brownish yellow
Ivory: various blues
Jadeite: green J.: very weak, whitish glow
Kornerupine: usually none;
  green K. from Kenia: yellow
Kunzite: strong; yellow-red, orange
Kyanite: weak, red
Labradorite: yellow striations
Lapus lazuli: strong; white
Leucite: orange or none
Magnesite: blue, green, white
Moonstone: weak; bluish, orange

Moss agate: varies
Morganite: weak, violet
Nepheline: bright; blue, weak; orange
Opal: white: bluish, brown, green
  black: usually none
  fire: greenish to brown
Pearls: weak
  natural black P.: red to reddish
  freshwater P.: strong, pale green
Pectolite: greenish yellow to yellow
Periclase: weak; orange
Petalite: weak, orange
Phenakite: pale greenish, blue
Phosgenite: yellow, orange-yellow
Phosphophyllite: violet
Pollucite: orange to pinkish
Prehnite: weak, orange
Rhodochrosite: weak; red
Rose quartz: weak; dark violet
Ruby: strong; carmine red
Sapphire: blue: violet or none
  yellow: weak; orange
  colorless: orange-yellow or violet-blue
Scapolite: pink: orange, pink
  yellow: violet, blue-red
Scheelite: strong, light blue
Serpentine: weak, greenish
Simpsonite: blue, bluish white, weak; yellow
Smithsonite: blue-white, pink, brown
Smoky quartz: usually none, rarely weak;
  brown-yellow
Sodalite: strong; orange
Sphalerite: yellow to orange, red
Spinel: red: strong; red
  blue: weak; reddish or green
  green: weak; reddish
Strontianite: white, olive green, blue-green
Taaffeite: various green
Thaumasite: white
Topaz: pink: weak; brown
  red: weak; brown-yellow
  yellow: weak; orange-yellow
Tourmaline: colorless: weak;
  green-blue
  pale yellow: weak; dark green
  red: weak; red-violet
  pink, brown, green blue: none
Tugtupite: bright to weak; orange
Turquoise: weak; greenish-yellow, light blue
Ulexite: green to yellow, blue
Variscite: pale green, green
Willemite: green
Witherite: blue, yellow-white, white
Wollastonite: blue-green
Zektzerite: light yellow
Zircon: blue: weak; light blue
  red and brown: weak; dark yellow

Inclusions

There is hardly a gemstone which is completely "clean." Most of the time, they contain foreign matter or some kind of dislocation or irregularity in the crystal lattice. Sometimes only under the microscope can these "flaws" be discovered. In the lingo of the trade, one does not speak of flaws, but rather of inclusions, because the term *flaw* has a derogatory meaning that is not justified.

Since inclusions are not accidental appearances but rather are subject to strict conformities with natural law, they can tell a lot about the origin of the gemstone deposits and, in addition to that, they help in their identification. Even though each gemstone has its individual inclusion history, there are nevertheless forms of inclusions which can be grouped together and can be associated with specific gemstone types and/or with gemstone imitations.

Inclusions of minerals are quite common, such as those of the same material (for instance, diamond in diamond) or of a foreign one (for instance, zircon in sapphire). Even if small, they provide a constructive picture of the formation of the surrounding crystal (the *host crystal*). Included minerals can be older than the host crystal, as they were just surrounded. They can also have been formed from a melt at the same time as the host crystal, which surrounded the smaller ones because its growth rate was greater. There are also mineral inclusions that are younger than the host crystal; these were formed out of liquids that entered into the crystal through cracks and fissures. In some cases (e.g. rutile needles in ruby or sapphire) inclusions form within the crystal as it cools.

Smoky quartz with star made of golden-yellow rutile needles; Minas Gerais/Brazil.

Amber with coal-like substance and insect inclusions; Samland, East Prussia/Russia.

Organic inclusions are only found in amber (see page 228 and the picture above). Parts of plants and insects are often preserved in it and bear witness to the life at the time, sometimes as much as 50 million years ago or more.

Irregularities in the crystal structure, marks of the crystallization phases, and color striations are all classed as inclusions. They can be formed by irregular growth from various crystallizing solutions.

Cavities also, when filled with liquid (water, liquid carbonic acid) or gases (carbon dioxide or monoxide), are also classed as inclusions. Where liquid and gas occur together, they are called two-phase inclusions. Liquid, gaseous, and small crystal inclusions are called three-phase inclusions. Completely empty cavities are not known. Air-filled bubbles are found frequently in obsidian, glass imitations, and synthetic gems, but not in mineral gems.

Even breaks and splits (called feathers), whether caused by internal stress or external pressure, are classed as inclusions. They can be observed in the interior of a stone, but they may also reach the surface. Air and solutions can enter a stone along such fissures and affect the color. If the stone is "heated," the foreign substance can be extruded, but scars show the old crack.

The trade and layman consider most inclusions as devaluing a stone because, under certain circumstances, color, optical properties, and mechanical resistance can be affected. However, some inclusions cause light phenomena that can produce the most valuable properties of some gems; for instance, cat's eye effect, asterism, silk (page 44), and dendrites (page 130). The golden inclusions of rutile in rock crystal or smoky quartz are most effective, especially when they form a star (page 49).

Only for diamonds do generally accepted standards for grading clarity exist (page 77). For all other gemstones, the individual speciman is judged for the quality effect of inclusions.

Natural emerald with oil residues that light up brightly in a natural crack (enlargement 40 ×).

Synthetic emerald from a Russian production. Characteristic V-structure (enlargement 20 ×).

Green glass emerald imitation with stars from small crystals, which came about through devitrification, as well as streaks, which are typical for glass (enlargement 35 ×).

51

Deposits and Production of Gemstones

Gemstones are found in many parts of the world, singly or grouped together. Groups large enough to be worked are called *deposits*. Places with a single find are simply called the location of discovery, place of discovery, or point of discovery. The work *occurrence* refers to all four terms.

The Finsch diamond mine, located in Northern Cape Province/South Africa, exploits a kimberlite pipe. Discovered in 1960, first put into production in 1965, now over 20 prospect levels deep.

Fluorite mine near Chiang Mai/Thailand. Without using machines, the strongly weathered rocks are being loosened, carried off, and sorted.

Types of Deposit

According to the way the gemstones have been formed, we differentiate between igneous or magmatic (formed out of the magma), sedimentary (formed by sedimentation), and metamorphic (re-formed from other rocks) deposits.

It is often more useful to speak of primary and of secondary deposits. That way, one differentiates between occurrences in which gemstones are still at the "first" location, i.e., at the place of the original formation, and those places to which they were transported to a "second" location.

In primary deposits the stones still have their original relationship with their host rock. The crystals are usually well preserved, but often the yield of the deposit is not very large. Many tons of *deaf* (nongem-bearing) rock have to be removed in the search for the gemstones.

In the case of secondary deposits, the gems have been transported from the place of their formation and deposited somewhere else. During this process, harder crystals become rounded, and softer ones are made smaller or even destroyed. According to the way the stones have been transported or according to the place of deposit, they are differentiated as *fluvial* (river), *marine* (in the sea), *littoral* (coastal), or *aeolian* (by the wind) deposits. Rivers can transport gem-bearing rocks many hundreds of miles. When the current diminishes and with it the transporting energy, the denser gems (for instance, diamond, zircon, garnet, sapphire, chrysoberyl, topaz, peridot, and

A gemstone mine in Sri Lanka, run by the government. The only machine run by an engine is the pump that sucks up the water entering into the mine.

tourmaline) are deposited before the lighter quartz sand. The gems left behind thus concentrate in certain places. This usually makes mining the deposit easier and more productive than working primary deposits.

Gemstone deposits which were deposited by river are called placers or alluvial deposits. Similarly, alluvial deposits can be found in the surf-pounded strands of the sea. In southwest Africa (Namibia), such diamond deposits are being worked very successfully (see photo, page 57). Small gem crystals can even be transported by the wind and enrich a particular place by sorting.

Seen genetically, between the primary and secondary deposits there are places where decomposition or weathering takes place. The gems are often found at the foot of steep cliffs or high mountains, and have collected in such places as part of the weathered debris; then, because the specifically lighter host rock was gradually carried away by rain and wind, the gems were left. These are called eluvial deposits.

Distribution of gems over the earth is irregular. Some regions are more favored than others: South Africa, south and southeast Asia, the Urals, Australia, Brazil, and the mountainous zones of the United States.

Mining Methods

Most gemstone deposits were discovered by accident. Even today systematic prospecting is primarily limited to diamond occurrences. The chief reason for this is that the diamond production is run by international companies and the price is controlled on a worldwide scale. This makes a large capital investment possible for prospecting.

Prospecting for nondiamondiferous deposits is usually accomplished by simple means, without modern techniques or scientific basis. The success of local prospectors in finding new deposits is surprising. A deposit being worked is called a mine.

With the exception of diamonds, mining methods in most countries are very primitive. In some regions, they have not changed in the last 2000 years (see picture on page 53). The increasing demand for gemstones has led some countries to modernize their prospecting. In the emerald mining of South Africa (see picture on page 92) and in the opal mines of Australia (see picture on page 56), for instance, there have been limited advances in ore processing. In Brazil at some places hydraulic mining is used for the sorting of the slope rubble.

The easiest way to collect gemstones is to discover stones at the surface, for instance, in a dry riverbed, in rock crevices, or in caves. But more often, the winning of gemstones takes a tremendous effort and a lot of hard work.

Gemstone mine in Ratnapura in Sri Lanka. With a simple windlass hoist, the earth containing gemstones, which is dug up deep down, is brought up for further processing.

55

Opal mining with quarryman; Lightning Ridge, New South Wales, Australia.

Extraction machines bring the mined rocks containing opal to the surface; Cooper Pedy/ South Australia.

Crystals grown into the mother-rock are loosened with hand tools, compressed-air (pneumatic) tools, or explosives. Underground mining has been done, to some extent, for centuries with both vertical shafts and transverse tunnels.

The mining of gemstones from secondary placer deposits is comparatively simple. With pickax and shovel the material containing gemstones is loosened, dug up, and carried off in baskets, to be worked at another location and examined for gemstones (see picture on page 54). In many countries where there are workable gemstone deposits, it is generally most cost-effective to employ unskilled labor rather than use expensive machinery. This is especially true in many developing nations.

If the secondary deposits containing gemstones are underneath a surface layer, the clay and sand layer is removed or shafts are built downwards. Such shafts, with minimal bracing, can be up to 30 ft (10 m) deep (see picture on page 55). The only modern piece of equipment in such gemstone mines are sump pumps to remove water which enters into the pit by rain or seepage.

Another way to exploit placer deposits is through prospecting in riverbeds. Various water-flow conditions are created through sluiceways and small dams for the purpose of selecting minerals. The less dense clay and sands are swept away, while the heavier gems remain. Workers agitate the material with long sticks to accelerate the sorting (see picture on page 85).

The actual separation of the gemstones from the gem-bearing material is done through another sorting out of the lighter minerals with the help of the moving water in step-like sluiceways and collection pits. Here, the gem-bearing material is further concentrated in baskets, which are washed, allowing lighter material to pan off over the edge of the baskets (see picture on page 89).

The final selection of a gemstone-quality material is always done by hand. The yield of gemstones is usually very low. Often only two or three little stones of gemstone quality are contained in the remaining concentration.

A special problem of gem production is the taking of gemstone-quality stones by the workers themselves, which can imperil every mine by undercutting economically viable prices. There is endless ingenuity employed to smuggle gems out of the mines, but, at the same time, methods for prevention are becoming more stringent. Diamond mines are the best protected. (See page 72 for more about the deposits and mining of diamonds.)

Diamond mine of Oranjemund/Namibia. Huge retaining barriers hold back the water of the Atlantic Ocean, so that the placers containing diamonds in the strip along the coast can be mined once a 100-ft. (30-m)-tall sand cover has been removed.

Cutting and Polishing of Gems

The oldest way of decorating a gemstone is the scratching of figures, symbols, and letters on it. From this, the art of stone engraving developed. The origins of gem cutting can be found in India. Up to about 1400, only the natural crystal faces or cleavage planes of transparent stones were polished in order to give them a higher luster and improved transparency. But even before then, opaque stones—mainly agates—were cut and polished with hard sandstone either flat or slightly arched (cabochon).

Stone cutting culminates in the faceted stone. There were reports of a faceted diamond in Venice as early as 800 A.D., but according to other opinions, the facet cut was only developed around the 15th century. For a long time the technique of faceting was kept a strict secret. (See also diamond cutting on page 80.)

Since the beginning of modern times, Amsterdam and Antwerp have developed as diamond cutting centers. Idar-Oberstein/Rhineland-Palatinate became the center of agate and colored stone manufacturing in the 16th century. Nowadays numerous cutting centers are being developed around the world. In order to encourage these centers, many countries have prohibited the export of rough stones.

In the manufacture of gemstones, one differentiates between engraving (page 59), working of agates (page 60), colored stone (page 61), ball or cylinder cutting (page 62), working of diamond (page 64), and the piercing of gemstones (page 63). Commercially, there is no strict division between these fields of work.

Rolling off in clay of a seal made from limestone. The figures depict fighters, animals, and other creatures. The roller-seal stems from the early third century A.D.; discovered in Syria.

58

Cutting of agate with diamond sawblade.

Fine grinding of agate on the sandstone wheel.

Engraving on Stones

The art of stone engraving or carving (also called glyptography) refers to the cutting of cameos and of intaglios, as well as to the creation of small objets d'art and other ornamental pieces.

The oldest stone engravings were cylinders engraved with symbols and figures which were used as seals or amulets (see picture on page 58). They were made in the ancient kingdoms of Sumeria, Babylon, and Assyria. The oldest figures are the scarabs of the Egyptians.

Stone engraving was practiced in ancient Greece and reached a high standard in ancient Rome. Even though the glyptography received renewed activity through heraldry in the Middle Ages, in general, development stagnated. Only with the Renaissance was the art of stone-cutting revived in Italy. Today, glyptography is appreciated as an art, especially due to its creative use of modern tools. In antiquity agate, amethyst, jasper, carnelian, and onyx were first used for stone engraving; gradually other gemstones were also used. Today, virtually all gemstones are engraved, including diamond. (For modern engraving, see page 142; for the technique, see page 144.)

Polishing of agate on slowly rotating wheels without or with only a little cooling fluid.

Cutting
and Polishing
Agate

Large and heavy stone blocks used to be split with hammer and chisel along fissures or other lines marked in the stone; nowadays they are almost without exception sawed with a diamond-charged circular saw. The blade is cooled with special coolants.

Agate is first roughly shaped on a carborundum wheel. To keep the stone steady, the cutter holds it between his knees. The wheel is cooled with water. The final shaping is obtained on a sandstone wheel. The cutter sits on a chair with a support for his chest (see picture on page 59). Grooves in the wheel also enable the stone to be cut en cabochon.

The final process is polishing, giving the stone luster and showing up the fine structural lines. Polishing is done on a slowly rotating cylinder or wheel of beech wood, lead, felt, leather, or tin with the help of a polishing powder such as chromium oxide, tripoli, or other pastes. No coolant is used in this process, so special care must be taken otherwise the stone could be damaged by the resultant heat.

Machines have now been developed which cut flat stones automatically. Cabochons can also be cut this way with the aid of a model. (For coloring of agates, see page 136. For history of agate polishing, see page 138.)

Cutting and Polishing Colored Stones

In the manufacturing process, the term "colored stones" refers to all gems with the exception of diamonds. (In Germany, they exclude agates as well.) The cutting of the colored stones is called lapidary work; the cutter is known as a lapidary. Most lapidaries specialize in a certain gem or stone group, so that consideration can be best given to the characteristics of the stone, such as depth of color, pleochroism, or hardness.

Circular saws with edges containing diamond powder instead of teeth are used first to cut the stone roughly to the required size. The final shape is given to the rough stone on a vertical, roughly grained carborundum wheel, cooled with water. Opaque stones, or those with inclusions, are cut en cabochon on grooved carborundum wheels.

Transparent stones, once roughly shaped, are faceted on horizontal grinding wheels. For this purpose, they are cemented with a special cement or shellac on one end of small 4-6 in (10-15cm)-long holders (putty sticks or "dops"). These putty sticks are guided at an angle, related to the facet being cut, by inserting the opposite end of the holder in the appropriate hole of a small board with rows of holes, fixed parallel or at right angles to the grinding wheel. Instead of such a board, a holder with a faceting head is used by amateurs to adjust the angles.

The substance from which the cutting wheel was made (lead, bronze, copper, tin, etc.) and the type of polishing powder (carborundum, diamond, titanium carbide), as well as the speed of the wheel, varies with the stone to be worked.

Prepolishing of colored stone at the vertically running, slightly moistened wheel.

Facetting on a horizontally rotating polishing wheel with the aid of a board that has holes for the gem holder.

Gemstone cutting with vertically running wheel and fiddle-bow drive; Sri Lanka.

The last process is polishing on a horizontally rotating wheel, or wooden cylinder, or on leather straps to remove the last traces of scratches and improve the luster. Melting processes at the stone surface and the formation of a very thin melting layer (the so-called Beilby layer) enhance the fine polishing effect.

In developing countries, gemstones are frequently cut en cabochon on hand-driven, vertically rotating lead or zinc wheels. Of course, with this method, a faceting that meets the requirement of exactness is not possible. Nevertheless, it is admirable the extraordinary skills the cutters use to form precious stones on these simple devices by which they bring to life many light figures of these stones.

Within recent years, automated cutting machines similar to those used for diamonds have been introduced for faceting colored stones. These produce fashioned gemstones of the standard calibrated dimensions increasingly demanded by the jewelry industry.

Firing, i.e., changing the color through applying heat to colored stones, will be discussed individually in connection with the respective gemstones.

Ball and Cylinder Cutting

Agates and colored stones are used for the making of stone balls and of irregularly cut so-called baroque stones with cylinders.

Ball Cutting
Small balls, up to about 3/4 in (20 mm) in diameter, can be made in larger numbers at the same time by hand or with partially automatic cutting machines (called ball mills). Larger stone balls can only be worked individually by hand. A hollowed-out, support wood piece guides vertically running cutting wheels; a cutting dop guides horizon-

tally turning cutting machines. Equal-sided cubes are normally the raw pieces used as the basic form for ball cutting.

Cylinder Cutting

The results of cylinder cutting (often called tumbling) are very different serendipitous stone shapes, which are called in the trade baroque stones. They are formed from little, irregular stone pieces in slowly rotating wood or steel cylinders. The stones polish each other through tumbling movements and added polishing substances of various grits. With machines running nonstop, this cutting procedure lasts between two and five weeks. Optimal end products can be achieved when the pieces being polished in the cylinders consist of raw pieces of the best quality of about the same size and the same hardness. (For forms of baroque stones, see page 252.)

Drilling Gemstones

Occasionally, gemstones have to be pierced in order to make jewelry. Formerly, this was done individually by hand with a fiddle-bow-driven drill. Today, mainly high-speed electrical drills are used. They have to be continuously provided with a coolant. Diamond powder and polishing paste aid the drilling procedure. One usually drills from both sides, so that the stone does not splinter at the original hole. In specialized companies, ultrasound drilling machines which work with vibration are used. The special advantage of these robotic devices is not only that the drilling time is shortened but that it is also possible to drill differently formed holes. (For more about the drilling of pearls, see page 237.)

Drilling of colored stone with pointy drill and fiddle-bow drive.

Cutting and Polishing Diamond

The following processes can be distinguished in the working of diamond: pre-examination, cleaving or sawing, bruting, cutting, and polishing.

Before the actual working of a diamond is started, the cutter first must study the "inner life" of a rough piece; he must get to know the structure of the crystal and judge the inclusions. That is done with a magnifying glass (loupe), and it can take several weeks in the case of more valuable pieces. Diamonds with a mat surface first must be pre-cut for this purpose; they receive, so to speak, a window through which one looks into the diamond. Once it is decided whether a stone, which is quite large, will remain at that size or whether it is better to divide it into several fragments, the intended cleaving or sawing direction is marked on the crystal with ink.

At one time larger diamonds were divided by cleaving. To do this, an expert prepares the diamond, then carefully positions a metal blade and strikes it with a mallet. The cleavage surfaces are always octahedral planes. The Cullinan diamond, the largest

Cleaving of diamond with wedge and with a blow by hand.

Sawing of diamond on a circular saw covered with diamond powder.

Two diamonds receive their raw shape for the brilliant cut by rubbing them against each other.

Faceting of diamond on horizontally rotating cutting wheel with the help of a dop.

diamond ever found—as large as a fist—was cleaved in 1908 in Amsterdam by the firm of Asscher, in the first stage into three parts, then again, so that finally nine large and 96 smaller brilliants could be cut from it.

Although the technique of cleaving is common, it often happened that a rough stone was broken because inner tensions and hidden cracks had not been recognized. Because of this, sawing diamonds has gradually become common practice since the turn of the century. A special advantage obtained from sawing diamonds is that a higher yield can be obtained from the rough stone. Well-formed octahedrons are sawn, for instance, through the central plane, or just above it, thus yielding a more favorable rough form for cutting brilliant. The sawing plane becomes the future table of the diamond-brilliant.

The disk of a diamond saw (about 4 in or 10 cm diameter) is made of copper, bronze, or another alloy, is paper-thin, and is impregnated with diamond dust, which has to be continually renewed. It rotates at about 4500 to 6000 revolutions per minute (r.p.m.). The rough stone is held in a vise-like arrangement. The sawing process is time-consuming: for a one-carat stone of ¼ to ¾ in (6-7 mm) diameter, about five to eight hours are needed. Lately, diamonds are also divided with laser beams. The advantage of this procedure is that one does not have to adhere to a structural direction of the crystal, unlike when cleaving or sawing.

The next process is bruting (or girdling), which gives the diamond its rough shape with girdle, crown, and pavillion. Two diamonds are used in bruting; one fixed in a small lathe, the other cemented onto a dopstick and held in the operator's hand. These are ground against each other so that the stone edges become rounded, corresponding to the double cone of the brilliant shape. Those diamonds which are not supposed to get the brilliant cut are ground on a disk that is covered with diamond dust. Polishing or sawing of diamond is only possible with other diamond, because diamond varies in hardness on different crystal faces and in different directions (see the sketch on page 20).

It is necessary to examine the diamond thoroughly in order to utilize the differences in hardness when cutting it. According to statistical probability, diamond powder contains grains lying in all directions, so that there are directions at all cutting angles. The powder can thus be used to cut the less hard directions of the diamond crystal, since those will be scratched/cut much more than the harder directions.

The technique of polishing requires much experience. The facets of the diamond, held in a dop, are polished on a horizontally running wheel, about 12 in (30 cm) in size, primed with diamond powder and oil, and rotating at about 2000 to 3000 r.p.m.

Traditionally, and still in much of the industry, the corners of all facets and placement of angles are judged by eye by experienced cutters without elaborate instruments, only using a loupe to gauge the cut. Since the 1970s automatic polishing machines have been increasingly used for stones up to about a half carat. Larger rough and fine stones continue to be cut by the old methods.

The loss of material during cutting of diamonds is very large, as high as 50 to 60 percent. In the case of the Cullinan, it was about 65 percent. The cut-off diamond powder is always captured and used for other cutting procedures.

The brilliant is finally polished on the same scaife, but on another ring with finer diamond powder. (For the history and development of diamond cutting, see page 80.)

Types and Forms of Cut

There is no general rule that can be applied to the various cuts of gemstones. Therefore, one finds in the literature conflicting attempts to classify them. Nevertheless, a distinction according to types of cut and forms of cut is quite possible.

Types of Cut
Based on the optical impression of the cut gemstones, three main groups, or types of cut, can be named: the faceted cut, the plain cut, and the mixed cut. (Compare with the drawings on the next page.)

Faceted cut   The faceted cut receives its character from a variety of small planes, the facets. It is applied mostly for transparent gems. Most faceted cuts can be divided into three basic types, brilliant cuts with mostly rhomboid (lozenge) and triangular facets in a radial pattern, step cuts with trapezoid or rectangular facets in concentric rows, and mixed facet cuts combining both brilliant- and step-type facets. (For more about faceted cuts, see pages 61, 65, and 80.)

Plain cut   The plain cut can be executed plain (level tablet), or arched as a cabochon or as a sphere. No facets interrupt the even stone surface. The plain cut serves mainly for the cutting of agate and other opaque gemstones. (For more about the practice of employing the plain cut, see pages 58, 60, and 62.)

Mixed cut   The mixed cut is a combination of the faceted and plain cut. On the upper (or lower) part, the gemstone carries facets; on the other part, it is smooth, roundish, or plain. Occasionally there are cutting forms followed in which the cut is "mixed" on the same side of the stone.

Forms of Cut
An abundance of forms is derived from the basic types of the gemstone cuts. They can be round, oval, cone-shaped, square (carré), rectangular (baguette), triangular, and multi-cornered.

In addition, there are many forms that imitate familiar shapes, such as olive, pear (drop or pendeloque), little ships (navette or marquise), heart, trapeze, or barrel. A number of purely fantasy cuts can hardly be overlooked, and new forms are increasingly created by designers. (A selection of the well-known forms of cuts are pictured on the front inside cover.)

Types of Cut for Gemstones

Top and side view

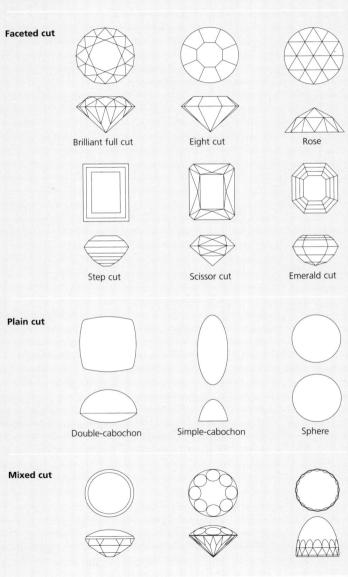

Faceted cut

Brilliant full cut

Eight cut

Rose

Step cut

Scissor cut

Emerald cut

Plain cut

Double-cabochon

Simple-cabochon

Sphere

Mixed cut

Description of Gemstones

In order to have a better overview of the large number of gemstones, it makes sense to group them according to their characteristics.

Scientific Classification
In the science of mineralogy, gemstones are classified according to their chemical compositions into eight classes. A ninth class refers to nonminerals, which may be noncrystalline organic combinations and/or rocks.

In the following table, the names of the gemstone species (e.g., corundum) as well as well-known varieties (e.g., ruby) are mentioned, in order to make it easier for laymen.

Mineral Classes

1. Elements   Diamond, gold, silver, sulfur

2. Sulfides   Chalcopyrite, greenockite, marcasite, proustite, pyrargyrite, pyrite, realgar, sphalerite, wurtzite, cinnabar

3. Halides   Boleite, creedite, fluorite, prosopite, sellaite, villiaumite

4. Oxides and hydroxides   Aschynite, alexandrite, anatase, bixbyite, brookite, cassiterite, chromite, chrysoberyl, corundum, cuprite, davidite, diaspore, euxenite, fergusonite, gahnite, gahnospinel, hematite, hercynite, hubnerite, ilmenite, magnetite, manganotantalite, midrolite, periclase, picotite, pleonaste, prase, psilomelane, pyrolusite, ruby, rutile, sapphire, senarmontite, simpsonite, spinel, stibiotantalite, taaffeite, tantalite, thorianite, wolframite, yttrotantalite, zincite

5. Nitrates, carbonates, borates   Ankerit, aragonite, azurite, barytocalcite, calcite, cerussite, chambersite, colemanite, coracite, coral, dolomite, gaylussite, gaspeite, hambergite, inderite, jeremejevite, kurnakovite, magnesite, malachite, mother of pearl, painite, parisite, pearl, phosgenite, rhodochrosite, rhodozite, shortite, siderite, sinhalite, smithsonite, stichtite, strontianite, ulexite, witherite

6. Sulfates, chromates, molybdates, wolframates   Anglesite, anhydritspar, barite, celestite, crocoite, gypsum, linarite, scheelite, wulfenite

7. Phosphates, arsenates, vanadates   Adamite, amblygonite, apatite, augelite, bayldonite, beryllonite, brazilianite, childrenite, cerulite, descloizite, durangite, eosphorite, goyazite, herderite, lazulite, legrandite, lithiophilite, ludlamite, mimetesite, monazite, montebrasite, phosphophyllite, purpurite, schlossmacherite, scorzalite, skorodite, triphylite, turquoise, vayrynenite, vanadinite, variscite, vivianite, wardite, zenotime

8. Silicates   Achroite, actinolite, agate, albite, almandite, amazonite, amethyst, analcime, andalusite, andesine, andrachite, anortite, anthophyllite, apophyllite, aquamarine, aventurine, aventurine feldspar, axinite, benitoite, beryl, bloodstone, blue quartz, bustamite, bytownite, canasite, cancrinite, carletonite, carnelian, catapleiite, chalcedony, charoite, chondrodite, clinohumite, clinozoisite, chloromelanite, chondrodite, chrysocolla, chrysoprase, clinozoisite, citrine, danburite, datolite, demantoid, diopside, dioptase, dravite, dumortiertite, precious beryl, ekanite, emerald, enstatite, epidote, eudialite, euclase, falcon's eye, feldspar, friedelite, gadolinite, garnierite, gold beryl, goshenite, garnet, grandidierite, grossularite, haüynite, helidore, hemimorphite, hessonite, hiddenite, hodgkinsonite, hornblende, howlite, hyalophane, hypersthene, idocrase, iolite, ilvaite, indicolite, jadeite, jasper, kornerupine, kunzite, kyanite, labradorite, lapis lazuli, lawnsonite, leifite, lepidolite, leucite, leucophane, meerschaum, melanite, melinophane, mesolite, milarite, moldavite, moonstone, morganite, moss agate, muskovite, natrolite, nepheline, nephrite, neptunite, obsidian, oligoclase, opal, orthoclase, palygorskite, pectolite, peridot, peristerite, petalite, petrified wood, phenakite, piemontite, pollucite, prasiolite, prehnite, pumpellyite, pyrope, pyrophyllite, pyroxmangite, quartz, rhodolite, rhodonite, rock crystal, rose quartz, rubellite, sanidine, sarder, scapolite, schorl, scolezite, serandite, serendibite, serpentine, shattuckite, sillimanite, smaragdite, smoky quartz, sodalite, sogdianite, spessartite, sphene, spodumene, spurrite, stellerite, staurolite, sugilite, talc, tanzanite, thalenite, thaumasite, thombartite, thomsonite, thulite, tiger's eye, topaz, topazolite, tourmaline, tremolite, tsavorite, tugtupite, uvarovite, verdelite, verdite, vlasovite, willemite, wollastonite, yugavaralite, zektzerite, zircon, zoisite

9. Nonminerals   Agalmatolite, alabaster, amber, ammonite, anyolite, jet, kakortokite, melite, nuummite, odontolite, onyx, tufa, unakite, whewellite

Commercial Classification
For practical reasons and to help the layman, the descriptions of gemstones of the world are arranged into five groups with similar characteristics rather than a strictly mineralogic classification.

I. Best-known gemstones   These include all the stones that have been traditionally traded and are otherwise generally known. These stones are set and worn as pieces of jewelry or worked into objets d'art.

The order within this main group is determined by Mohs' hardness.

This group comprises diamond to malachite, pages 70 to 177.

II. Lesser-known gemstones   Most of the representatives of this second group had been rare in the trade until recently, but are becoming more popular today. Formerly appreciated mostly by collectors, they are today made into jewelry and worn just like any other better-known gemstone.

The sequence within this main group is also determined by Mohs' hardness, because it gives an indication of their practical suitability as gemstones.

The representatives of this group are described on pages 178 to 203.

III. Gemstones for collectors   There are many minerals which are only cut by collectors, amateur or professional lapidaries, although they have no practical use in jewelry, because they are too soft, too brittle, or too rare. Gem lovers collect such faceted or en cabochon stones as rarities or as small objets d'art.

A selection of these gemstones is presented on pages 204 to 217.

IV. Rocks as gemstones   This group (pages 218 to 223) is considered in the fringe zone of gems. As fashion or costume jewelry, rocks (mineral aggregates in large combinations) are becoming increasingly important. Striking structures and beautiful colors of such rocks can be seen in the gemstone trade.

V. Organic gemstones   This group includes materials which, even though they are of organic origin, have preserved or acquired a certain stone character. They are an important part of the trade, especially with respect to amber or pearl. Many are made up of minerals; pearls are mostly aragonite, bone is largely apatite.

Representatives of organic gemstones are on pages 224 to 239.

Best-Known Gemstones

This group includes all those stones that have been traditionally represented in the trade or which are otherwise generally known. (Compare also to page 69.)

Diamond

Color: Colorless, yellow, brown, rarely green, blue, reddish, orange black
Color of streak: White
Mohs' hardness: 10
Density: 3.50–3.53
Cleavage: Perfect
Fracture: Conchoidal to splintery
Crystals: (Cubic), mainly octahedrons, also cubes, rhombic dodecahedrons, twins, plates
Chemical composition: C, crystallized carbon

Transparency: Transparent to opaque
Refractive index: 2.417–2.419
Double refraction: None
Dispersion: 0.044(0.025)
Pleochroism: None
Absorption: Colorless and yellow D.: 478 465, 451, 435, 423, 415, 401, 390; Brown and greenish D.: (537),504, (498)
Fluorescence: Very variable; Colorless and yellow D.: mostly blue Brown and greenish D.: often green

The name diamond refers to its hardness (Greek—*adamas*, the unconquerable). There is nothing comparable to it in hardness. Its cutting resistance is 140 times greater than that of ruby and sapphire, the gemstones next in hardness after diamond. However the hardness of a diamond is different in the individual crystal directions. This allows one to cut diamond with diamond and/or with diamond powder. Because of the perfect cleavage, care must be taken not to accidently bang against an edge of a diamond, and also when setting it. Its very strong luster sometimes enables the experienced eye to differentiate between a diamond and its imitations.

Diamond is generally insensitive to chemical reactions. High temperatures, on the other hand, can induce etchings on the facets. Therefore special care must be taken during soldering!

In the last fifty years, it has been recognized that there are various types of diamonds with different characteristics. Science differentiates between type Ia, Ib, IIa, and IIb. This is of little importance to the trade, but does assist the cutter. Due to the optical effects, the high hardness, and its rarity, the diamond is considered the king of gemstones. It has been used for adornment since ancient times.

See the following for further details:

Diamond cutting, page 64
Diamond deposits, page 72
Diamond trade, page 74
Quality evaluation, page 76

Famous diamonds, page 78
Development of the brilliant cut, page 80
New diamond imitations, page 242
Testing for genuineness, page 246

1 Diamond, brilliant 0.49ct
2 Diamond, marquise 0.68ct
3 Diamond, brilliant 2.22ct
4 Diamond, 3 baguettes, total 0.59ct
5 Diamond, brilliant 0.21ct
6 Diamond, 2 old cut 0.97ct
7 Diamond, 2 brilliants 0.57ct
8 Diamond, brilliant 2.17ct
9 Diamond, 3 roses, total 0.67ct
10 Diamond, 2 marquises, total 0.69ct
11 Diamond, twice sawn, total 1.43ct
12 Diamond, 10 brilliants
13 Diamond, 9 brilliants
14 Diamond, white rough, total 6.37ct
15 Diamond, colored rough, total 10.22ct
16 Diamond aggregate, 8.26ct
17 Diamond crystal on kimberlite
18 Diamond crystal, 8.14ct

Figs. 1–15 twice the original size; Figs. 16–18 reduced by one-third in size.

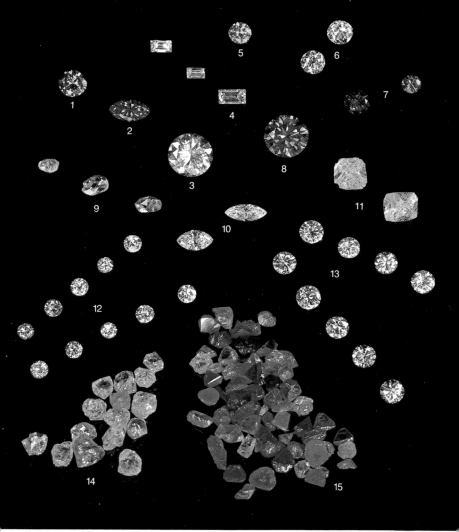

originally exclusively done by hand, is done today almost fully automatically. The tendency of diamonds to adhere to grease is also utilized. Diamond is, in contrast to other materials, hard to moisten, i.e., even in water it actually does not get wet. Therefore, from the concentrate, which glides over grease vibrating tables, the diamonds get stuck to the grease, while other minerals continue to glide along.

Other diamond sorting methods include electrostatic separation, with optical selection by photo-cells, or by using X rays or ultraviolet rays, utilizing the fluorescence.

Despite the usage of such modern techniques, the final selection must be done by hand, because other minerals end up in the final concentrate besides diamonds.

The diamond content of the host rocks varies, depending on the occurrences. It is usually higher in placers than in kimberlite pipes. In some placer deposits, a yield of ⅙ carat per cubic meter of rock is still economic. However, diamond content is misleading. Australia's Argyle mine (the world's largest diamond mine) is hosted by a lamproite and produces a high yield. Since diamonds are individually valued, the degree of size and quality of stones can dramatically affect the value such that yields a defined in number of carats per cubic meter or penton are not very informative.

Diamond Trade

About 80 percent of the world diamond production and rough supply is managed, i.e., controlled, by an enterprise, which is known by the short name "De Beers." Numerous splinter associations of a densely intertwined mammoth enterprise, trade associations as well as institutions and companies outside of the diamond trade, all hide behind the term "De Beers," for instance, "De Beers Consolidated Mines Limited," "DeBeers Centenary AG," "Central Selling Organisation," abbreviated CSO.

Derogatorily, one sometimes speaks of a diamond syndicate. It is factually correct, however, to call De Beers a monopoly, if one wants to indicate the concentration of power within the diamond trade.

All sorted and valued diamonds go to London to the CSO. Here, they are put together into lots (parcels) for the purpose of selling and are offered for a fixed price. Partial purchase does not exist, only by lot.

Only a few diamond cutters and diamond traders, who are accredited at the CSO, the "sightholders," are entitled to buy (currently there are 160 such individuals). They purchase the merchandise, which was put together by the CSO, at ten annual selling events, the "sights," in London, Lucerne, and Johannesburg, according to the registered wishes of the customers. The CSO delivers only rough stones.

The resale and the dividing of the lots is done by the "direct buyers," preferably on diamond exchanges (sometimes called diamond clubs or bourses). Such exchanges exist in Antwerp, Amsterdam, New York, and Ramat Gan in Israel, in addition to those in Johannesburg, London, Milan, Paris, and Vienna, and, since 1974, also in Idar-Oberstein/Rhineland-Palatinate. By far the most important are the four diamond exchanges in Antwerp, the largest diamond-trading location in the world. Diamond exchanges are not speculation exchanges in the usual sense of the word, but rather are large diamond markets.

De Beers also controls and influences through its selling system the prices of the brokers. The goal is to keep the value of the diamond high and to call a halt to any dubious maneuvers in the diamond trade.

Such collaboration among the leading production countries under the organized leadership of De Beers can be said to be in the interest of everyone, not just the diamond producers, but even, perhaps, the customer, the buyer of the diamond jewelry, because through this cartel prices are kept high and extreme variations in price are avoided.

Weekly production of a diamond mine in Namibia, about 30,000 carats.

Certain problems for the controlled market situation arise because some countries (currently Angola, Zaire, and Russia) do not have their diamond production or sale restricted through the cartel. Thus any diamonds which are freely offered at whatever rate a more or less free market will support can influence the prices throughout the world beyond the control of the organization. Nevertheless, diamonds have survived as a dependable investment through all political and economic ups and downs of recent decades. Through this, not only capital was secured, but also millions of jobs, directly or indirectly connected with the diamond trade, were preserved. After all, the market value of diamond production is over 90 percent of the entire gemstone trade.

Diamond Imitations The fact that the diamond can be confused, in appearance, with many gemstones, can lead to fraud, although not in the legitimate retail trade. A colorless diamond looks similar to rock crystal (page 116), precious beryl (page 96), cerussite (page 200), sapphire (page 86), scheelite (page 196), sphalerite (page 200), topaz (page 102), and zircon (page 108). Also many yellowish stones can look like diamonds to the eye of the layman.

Apart from these, there are various synthetic stones (compare to pages 242 and 243) that are used to imitate diamonds. Especially strontium titanate (fabulite), YAG, GGG (galliant), linobate, and cubic zirconia (CZ, djevalite) must be mentioned. A well-known diamond imitation made from glass is the so-called strass (page 242).

In 1970 the first gem-quality diamonds were synthesized, but these still do not compete with natural diamonds; so far they serve mainly scientific purposes. In the trade one also finds natural diamonds, colored artificially by various irradiation treatments.

Diamond-doublets are made with: upper part-diamond; lower part-synthetic colorless sapphire, rock crystal, or glass. Other doublets have synthetic spinel as upper part, and strontium titanate underneath.

Valuation of Diamonds

All diamonds that De Beers brings on the market are classified beforehand into one of the 5,000 different standards of quality, according to form, quality, color, and size. Besides London, there are such sorting centers in Lucerne in Switzerland, Gaborone in Botswana, Windhoek in Namibia, and Kimberley in South Africa.

Formerly, 20 percent of all diamonds were considered suitable for jewelry, having been of "gemstone quality." The rest were sold to the industry as so-called industrial diamonds to be used for drilling crowns, milling machines, cutting wheels, etc. Since 1983, another 20 percent have been classified as "almost gemstone quality" and are cut mainly in India. The smallest diamonds and diamonds of lesser quality can be worked and offered to buyers other than the strictly controlled diamond market.

In the valuation of faceted diamonds, color, clarity, cut, and carat are taken into consideration. These four c's decide the value of a diamond.

Grading for Color Diamonds are found in all colors. Mostly they are yellowish. In the grading, these are evaluated together with the purely colorless diamonds. The rarer strong colors (green, red, blue, purple, and yellow), the so-called fancy colors, are valued individually and fetch collector's prices. Brown and black diamonds also occur.

Formerly, terms and definitions in grading for color were not uniform and often confusing, until an international agreement was reached to cover the so-called "yellow series." This was published in 1970 as RAL 560 A5E. Since then, various institutions have come up with improved guidelines, especially the Gemological Institute of America (GIA), the International Diamond Council (IDC), and the Confédération Internationale de la Bijouterie, Joaillerie, Orfèvrerie des Diamants, Perles, et Pierres (CIBJO). Today, the IDC regulations, written in English, are accepted worldwide. In the United States, the GIA system is most commonly used.

The old grading terms, the "Old Terms," should not be used anymore. In fact, however, they are still in usage in the gemstone trade.

Experts use standard sample collections for consistent, comparison color grading.

Color Grading of Faceted Diamonds

CIBJO	IDC	GIA	Old Terms	RAL 560 A5E
very fine white+	exceptional+ white	D	River	blue-white
very fine white	exceptional white	E		
fine white+	rare white+	F	Top Wesselton	fine white
fine white	rare white	G		
white	white	H	Wesselton	white
slightly tinted white	slightly tinted white	I	Top Crystal	weakly tinted white
		J		
tinted white	tinted white	K	Crystal	tinted white
		L		
tinted 1	tinted color 1	M	Top Cape	weakly yellow
		N		
tinted 2	tinted color 2	O	Cape	yellowish
		P		
tinted 3	tinted color 3	Q	Light Yellow	weakly yellow
		R		
tinted 4	tinted color 4	S–Z	Yellow	yellow

Grading for Clarity   In Germany only the inner perfection is understood as "clarity," while in the United States and in Scandinavia aspects of the quality of the outer finish are taken into consideration. Enclosed minerals, cleavages, and growth lines affect clarity; they are collectively called inclusions, but formerly were called "flaws" or "carbon spots." Polished diamonds without any inclusions under a 10 × loupe are considered "flawless." Inclusions visible with larger magnification are not taken into account for grading.

Grading for Clarity of Faceted Diamonds

CIBJO	Definition	GIA
Lr flawless	Free of inclusions under 10 × magnification and absolutely transparent	If internally flawless
VVS very very small inclusions	Very few, very small inclusions, under 10 × magnification very difficult to see	VVS 1 very very VVS 2 slightly included
VS very small inclusions	Very small inclusions, under 10 × magnification difficult to find	VS 1 very slightly included VS 2 slightly included
SI small inclusions	Small inclusions, easily recognized under 10 × magnification	SI 1 slightly included SI 2 slightly included
PI distinct inclusions	Inclusions, immediately recognized under 10 × magnification, but not diminishing brilliance	I1 included I
PII larger inclusions	Larger and/or numerous inclusions slightly diminishing the brilliance recognizable with the naked eye	I2 included II
PIII large	Large and/or numerous inclusions, diminishing the brilliance considerably	I3 included III

According to CIBJO, it is permitted to subdivide the clarity grades VVS, VS, and SI into two subgroups each, for stone sizes over 0.47 ct. Definitions assume a trained grader working under favorable conditions.

Grading for Cut   To grade for cut, the type and shape of cut, proportions, and symmetry as well as outer marks are taken into consideration. In Germany the normal cut is the "fine brilliant cut" (page 81), in the rest of Europe the "Scandinavian-standard brilliant." In the United States, the only widely used cut grading system is that of the American Gem Society, based on the "AGS Ideal Cut." The following table shows the terms and definitions for grading in a simplified form, according to the RAL 560 A5E, published in 1971.

Grading of Cut of the Diamond Brilliants

RAL 560 A5E	Definition
Very good	Exceptional brillance. Few and only minor outer marks. Very good proportions.
Good	Good brilliance. Some outer marks. Proportions with some deviations.
Medium	Slightly less brilliance. Several larger outer marks. Proportions with considerable deviations.
Poor	Brilliance considerably less. Larger and/or numerous outer marks. Proportions with very distinct deviations.

Famous Diamonds

A number of diamonds are well known and famous because of their size, beauty, or their adventurous past. (Glass replicas shown opposite.)

1 **Dresden**   Green, 41 carats. Probably from India; early history not known. Supposedly bought in 1742 by Friedrich August II, Duke of Saxony, for 400,000 taler. Kept in the Green Vaults in Dresden, Germany.

2 **Hope**   45.52 carats. Appeared 1830 in the trade and was bought by the banker H. Ph. Hope (hence the name). Probably recut from a stolen stone. Since 1958 in the Smithsonian Institution, Washington, D.C.

3 **Cullinan I, or Star of Africa**   530.20 carats. Cut from the largest rough gem diamond ever found of 3106 carats, the Cullinan (named after Sir Thomas Cullinan, chairman of the mining company) together with 104 other stones, by the firm Asscher in Amsterdam in 1908. Adorns the sceptre of the English king's insignia. Kept in the Tower of London; largest cut diamond of fine quality.

4 **Sancy**   55 carats. Said to have been worn by Charles the Bold around 1470. Bought 1570 by Seigneur de Sancy (hence the name) from the French ambassador to Turkey. Now on display at the Louvre in Paris.

5 **Tiffany**   128.51 carats. Found in Kimberley mine, South Africa, in 1878; rough weight 287.42 carats. Bought by the jewelers Tiffany in New York and cut in Paris with 90 facets.

6 **Koh-i-Noor**   108.92 carats. Originally a round stone of 186 carats belonging to the Indian Raj. Plundered in 1739 by the Shah of Persia, who called it "Mountain of Light" (*Koh-i-Noor*). Came into possession of the English East India Company, which presented it to Queen Victoria in 1850. Recut, it was set in the crown of Queen Mary, wife of George V, and then in a crown made for the mother of Queen Elizabeth II; now in the Tower of London.

7 **Cullinan IV**   63.60 carats. One of the 105 cut stones from the largest gem diamond ever found, the Cullinan (see 3 above). Also in the crown of Queen Mary; can be removed from this and worn as a brooch. Kept in the Tower of London.

8 **Nassak**   43.38 carats. Originally over 90 carats and in a Temple of Shiva near Nassak (hence the name) in India. Looted in 1818 by the English; recut 1927 in New York. Acquired by the King of Saudi Arabia in 1977.

9 **Shah**   88.70 carats. Came from India, shows cleavage planes, partially polished. Has three inscriptions of monarchs' names (amongst them the Shah of Persia's— hence the name). Given in 1829 to Tsar Nicholas I. Kept in the Kremlin, Moscow.

10 **Florentine**   137.27 carats. Early history steeped in legend. In 1657 in the possession of the Medici family in Florence (hence the name). During the 18th century in the Habsburg crown, then used as brooch. Whereabouts after World War I are unknown.

Other important cut diamonds (in carat): Cullinan II. (317.40, picture on page 9), Centenary (273.85), De Beers (234.50), Great Mogul (280.00), Jonker (125.35), Jubilee or Reitz (245.35), Nizam (277.00), Orloff (189.60), Regent or Pitt (140.50), Victoria (184.50).

The greatest rough stones found suitable for gem purposes (in carats): Cullinan (3106), Excelsior (995.2), Star of Sierra Leone (968.9), Incomparable (890), Great Mogul (787.5), Woyie River (770), President Vargas (726.6), Jonker (726), Jubilee (or Reitz 650.8), Dutoitspan (616), Baumgold (609), Lesotho (601.25), Centenary (599.0), Nizam (440.0), De Beers (428.5).

Corundum Species

Two color varieties of corundum are used for making jewelry, the red ruby and the sapphire which comprises all other colors (page 86). Common corundums, i.e., those not of gemstone quality, serve as cutting and polishing material. The well-known polishing material emery is mainly fine-grain corundum, to which magnetite, hematite, and quartz are added. The name corundum has its origin in India and probably referred to ruby.

Ruby   Corundum Species

Color: Varying red
Color of streak: white
Mohs' hardness: 9
Density: 3.97–4.05
Cleavage: None
Fracture: Small conchoidal, splintery, brittle
Crystal system: (Trigonal) hexagonal prisms or tables, rhombohedrons
Chemical composition: Al_2O_3, aluminum oxide

Transparency: Transparent to opaque
Refractive index: 1.762–1.778
Double refraction: —0.008
Dispersion: 0.018 (0.011)
Pleochronism: Strong; yellow-red, deep carmine red
Absorption: 694, 693, 668,659, 610–500, 476, 475, 468
Fluorescence: Strong: carmine red

Ruby is thus named because of its red color (Latin—*ruber*). It was not until about 1800 that ruby, as well as sapphire, was recognized as belonging to the corundum species. Before that date, red spinel and the red garnet were also designated as ruby.

The red color varies within each individual deposit, so it is not possible to determine the source area from the color. The designations "Burma ruby" or "Siam ruby" are therefore strictly erroneous, and refer more to quality than origin. The most desirable color is the so-called "pigeon's blood," pure red with a hint of blue. The distribution of color is often uneven, in stripes or spots. The substance that provides the color is chromium, and in the case of brownish tones, iron is present as well. As a rough stone, ruby appears dull and greasy, but, when cut, the luster can approach that of diamond. Heat treatment is commonly used to improve the color.

Ruby is the hardest mineral after diamond. However, the hardness varies in different directions. Ruby has no cleavage, but has certain preferred directions of parting. Because of brittleness, care must be taken when cutting and setting.

Inclusions are common. They are not always indicative of lower quality, but show the difference between a natural and a synthetic stone. The type of inclusion (minerals, growth structures, canals, or other cavities) often indicates the source area.

Included rutile needles bring about either soft sheen (called silk) or, when cut en cabochon, the rare cat's eye effect (see page 83, no. 5), which shimmers over the surface of asterism—a six-rayed star (see page 83, no. 4), which shimmers over the surface of the stone when it is moved. Nowadays there are also Trapiche Rubies on the market. Their appearance is equal to the Trapiche Emeralds (page 92).

---

1 Ruby, 5 faceted stones
2 Ruby, 2 drops, 2.51ct, Thailand
3 Ruby, engraved cabochon, 30.97ct
4 Star ruby,
5 Ruby cat's eye, 6.64ct

6 Ruby, 4 tabular crystals
7 Ruby, 3 prismatic crystals
8 Ruby, rolled crystal
9 Ruby, tabular crystal
10 Ruby in host rock, Sri Lanka

Deposits   The host rocks of ruby are metamorphic dolomite marbles, gneiss, and amphibolite. The yield of rubies from such primary deposits is not economically profitable. Rather, secondary alluvial deposits are worked. Because of its high density, ruby is normally separated through the washing of river gravels, sands, and soil, then concentrated, and finally picked out by hand.

Production methods are still as primitive as they were a hundred years ago in many locations. In state-owned mines, on the other hand, the usage of machinery is not exactly the rule, but much more frequent than in private companies. Some state-regulated companies (e.g., Mogok in Burma) lately even work with highly mechanized machinery both above- and underground.

Some of the most important deposits are in Burma (Myanmar), Thailand, Sri Lanka, and Tanzania. For centuries, the most important have been in upper Burma near Mogok. The ruby-bearing layer runs several yards under the surface. Apparently only about one percent of production is of gem quality. Some of the rubies are of pigeon's blood color. They are considered to be the most valuable rubies of all. Large stones are rare. Minerals found together with ruby, often also of gemstone quality, are precious beryl, chrysoberyl, garnet, moonstone, sapphire, spinel, topaz, tourmaline, and zircon. In the early 1990s large new deposits were discovered at Mong Hsu in Myanmar. Rubies from Thailand often have a brown or violet tint to them. They are found southeast of Bangkok in the district of Chantaburi in clayey gravels. Shafts are sunk to a depth of 26 ft (8 m). However, in recent years Thai ruby production has been declining.

In Sri Lanka deposits are situated in the southwest of the island in the district of Ratnapura. Rubies from these deposits (called *illam* by the local population) are usually light red to raspberry-red. Some of the rubies are recovered from the river sands and gravels.

Since the 1950s Tanzania has produced a decorative green rock, a zoisite (any-olite), with quite large, mostly opaque rubies (page 161, numbers 12 and 14). Only a few crystals are cuttable, most being used as decorative stones. On the upper Umba River (northwest Tanzania), on the other hand, rubies with gemstone quality have been found that are violet to brown-red.

Other significant producers of rubies include Afghanistan, Cambodia, Kenya, Madagascar, and Vietnam. There are also less important deposits in Australia, Brazil, India, Malawi, Nepal, Pakistan, the United States, and Zimbabwe.

Famous Rubies   Ruby is one of the most expensive gems, large rubies being rarer than comparable diamonds. The largest cuttable ruby weighed 400ct; it was found in Burma and divided into three parts. Famous stones of exceptional beauty are the Edwardes ruby (167ct) in the British Museum of Natural History in London, the Rosser Reeves star ruby (138.7ct) in the Smithsonian Institution in Washington, D.C., the De Long star ruby (100ct) in the American Museum of Natural History in New York, and the Peace ruby (43ct), thus called because it was found in 1919 at the end of World War I.

Many rubies comprise important parts of royal insignia and other famous jewelry. The Bohemian St. Wenzel's crown (Prague), for instance, holds a nonfaceted ruby of about 250ct. But some gems, thought to be rubies, have been revealed as spinels, such as the "Black Prince's ruby" in the English State crown (page 9) and the "Timur Ruby" in a necklace among the English crown jewels. The drop-shaped spinels in the crown of the Wittelsbachs dating from 1830 were also originally thought to be rubies.

Prospecting in the gemstone-bearing bed of a river; Sri Lanka.

Working Today rubies are often cut in the countries where they were found. As the cutters usually aim for maximum weight, the proportions are not always satisfactory, so that many stones have to be recut by dealers in other countries. Transparent qualities are cut in step and brilliant cut; less transparent stones, en cabochon or they are formed to carvings. Only synthetic rubies are used for watches and bearings, formerly the most important technical application for natural stones.

Possibilities for Confusion With almandite (page 104), pyrope (page 104), spinel (page 100), topaz (page 102), tourmaline (page 110), and zircon (page 108).

Since the beginning of the 20th century, there have been synthetic rubies with gemstone quality (page 243); these resemble natural ones especially in their chemical, physical, and optical properties. But most of them can be recognized by their inclusions as well as by the fact that they, in contrast to natural rubies, transmit shortwave ultraviolet light.

Numerous imitations are on the market, especially glass imitations and doublets. These have a garnet crown and glass underneath, or the upper part is natural sapphire with synthetic ruby underneath. There are many false names in the trade such as Balas ruby (= spinel), Cape ruby (= pyrope), and Siberian ruby (= tourmaline).

Color: Blue in various tones, colorless, pink,
    orange, yellow, green, purple, black
Color of streak: White
Mohs' hardness: 9
Density: 3.95–4.03
Cleavage: None
Crystal system: (Trigonal), doubly pointy,
    barrel-shaped, hexagonal pyramids,
    tabloid-shaped
Chemical composition: Al_2O_3 aluminum
    oxide
Transparency: transparent to opaque
Refractive index: 1.762–1.788

Double refraction: —0.008
Dispersion: 0.018 (0.011)
Pleochroism: Blue: definite; dark blue,
    green-blue
    Yellow: weak; yellow, light yellow
    Green: weak; green-yellow, yellow
    Purple: definite; purple, light red
Absorption spectrum:
    Blue, Sri Lanka, <u>471</u>, <u>460</u>, 455, <u>450</u>, 379
    Yellow, <u>471</u>, <u>460</u>, 450
    Brown, 471, 460–450
Fluorescence: Blue S.: none;
    Colorless S.: orange-yellow, violet

The name sapphire (Greek—blue) used to be applied to various stones. In antiquity and as late as the Middle Ages, the name sapphire was understood to mean what is today described as lapis lazuli. Around 1800 it was recognized that sapphire and ruby are gem varieties of corundum. At first only the blue variety was called sapphire, and corundums of other colors (with the exception of red) were given special, misleading names, such as "Oriental peridot" for the green variety or "Oriental topaz" for the yellow type.

Today corundums of gemstone quality of all colors except red are called sapphire. Red varieties are called rubies (page 82). The various colors of sapphire are qualified by description, e.g., green sapphire or yellow sapphire. Colorless sapphire is called leuko-sapphire (Greek—white), pinkish orange sapphire Padparadscha (Sinhalese for "Lotus Flower").

There is no definite demarcation between ruby and sapphire. Light red, pink, or violet corundums are usually called sapphires, as in this way they have individual values in comparison with other colors. If they were grouped as rubies, they would be stones of inferior quality. The coloring agents in blue sapphire are iron and titanium; and in violet stones, vanadium. A small iron content results in yellow and green tones; chromium produces pink, iron, and vanadium orange tones. The most desired color is a pure cornflower-blue. In artificial incandescent light, some sapphires can appear to be ink-colored or black-blue.

Through heat treatment at temperatures of about 3100–3300 degrees F (1700–1800 degrees C), some cloudy sapphires, nondistinct in color, can change to a bright blue permanent color.

Hardness is the same as ruby and also differs clearly in different directions (an important factor in cutting). There is no fluorescence characteristic for all sapphires.

Inclusions of rutile needles result in a silky shine; oriented, i.e., aligned, needles cause a six-rayed star sapphire (page 87, no. 2).

---

1 Sapphire, oval, 5.73ct, Thailand
2 Star sapphire, 9.46ct, Burma
3 Sapphire, brilliant cut, 2.81ct
4 Sapphire, 6 faceted stones,
    together 2.34ct, Tanzania
5 Sapphire, oval, 1.62ct, Sri Lanka
The illustrations are 30 percent larger than the originals.

6 Sapphire, drop, 6.09ct
7 Sapphire, yellow,
    11.32ct, Sri Lanka
8 Sapphire, antique cut, 5.18ct
9 Sapphire, antique cut, 3.74ct
10 Sapphire, 5 crystal shapes

Beryl Species

Several color varieties of beryl are used as gemstones. Deep green beryls are called emerald, blue aquamarine (page 94). All beryls of other colors of gemstone quality are called precious beryl (page 96). The name beryl comes from India and has always been associated with the gemstone.

Emerald

Beryl Species

Color: Emerald green, green, slightly yellowish-green
Color of streak: White
Mohs' hardness: 7½–8
Density: 2.67—2.78
Cleavage: Indistinct
Fracture: Small concoidal, uneven, brittle
Crystal system: (Hexagonal), hexagonal prisms
Chemical composition: $Al_2Be_3Si_6O_{18}$

aluminum beryllium silicate
Transparency: Transparent to opaque
Refractive index: 1.565–1.602
Double refraction: –0.006
Dispersion: 0.014 (0.009–0.013)
Pleochroism: Definite; green, blue-green to yellow-green
Absorption spectrum: <u>683</u>, <u>681</u>, 662, 646, <u>637</u>, (606), (594), <u>630–580</u>, 477, 472
Fluorescence: Usually none

The name emerald derives from Greek *smaragdos*. It means "green stone" and, in ancient times, referred not only to emeralds but also probably to most green stones.

Emerald is the most precious stone in the beryl group. Its green is incomparable, and is therefore called "emerald green." The coloring agent for the "real emerald" is chrome. Beryls that are colored by vanadium ought to be called "green beryl" and not emerald. The color is very stable against light and heat, and only alters at 1292–1472 degrees F (700-800 degrees C). The color distribution is often irregular; a dark slightly bluish green is most desired.

Only the finest specimens are transparent. Often the emerald is clouded by inclusions (compare to page 49). These are not necessarily classified as faults, but are evidence as to the genuineness of the stone as compared with synthetic and other imitations. The expert refers to these inclusions as *jardin* (French—garden).

The physical properties, especially the density, refractive index, and double refraction, as well as the pleochroism, vary according to source area. All emeralds are brittle and combined with internal stress, sensitive to pressure; care must be taken in heating them. They are resistant to all chemicals which are normally used in the household.

Deposits Emeralds are formed by hydrothermal processes associated with magma and also by metamorphism. Deposits are found in biotite schists, clay shales, in limestones, with pigmatites. Mining is nearly exclusively from host rock, where the emerald has grown into small veins or on walls of cavities. Alluvial placers are very unlikely to come about as the density of emerald is near that of quartz. Therefore, rare secondary deposits are mostly formed by weathering.

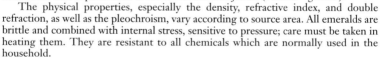

1 Emerald in host rock
2 Emerald, oval, 0.91ct, Colombia
3 Emerald, 2 pear shapes, 1.59ct
4 Emerald, 2 emerald cuts
5 Emerald, antique cut, 4.14 ct, South Africa
6 Emerald, oval, 1.27ct
7 Emerald, cabochon, 5.24ct
8 Emerald, cabochon, 4.26ct
9 Emerald, cabochon, 3.11ct
10 Emerald, crystal, Brazil

Work done in a sorting plant of an emerald mine; Cobra mine/South Africa

Significant deposits are in Colombia, especially the Muzo mine northwest of Bogotá. First mined by native tribes, the Muzo deposit was abandoned and rediscovered in the 17th century. The mine yields fine-quality stones of a deep green color. Mining, apart from shafts, is mainly by step-form terraces. The emerald-bearing, soft broken rock is loosened with sticks, lately also through blasting or with bulldozers, and the emeralds picked out by hand. The host rock is a dark carbonaceous limestone. Accompanying minerals are albite, apatite, aragonite, barite, calcite, dolomite, fluorite, and pyrite.

Another important Colombian deposit, the Chivor Mine, is northeast of Bogotá. It was also mined by Native Americans. The host rock is gray-black shale and gray limestone. It is mined in terraces, and also from shafts.

During recent decades further emerald deposits, which promise to be successful, have been found in Colombia. There is always a high demand for the rare, so-called Trapiche emeralds found exclusively in Colombia, a wheel-like growth of several prismatic crystals. Only a third of the Colombian emeralds is worth cutting. Stones larger than nut size are usually low quality or broken.

In Brazil there are various deposits in Bahia, Goias, and Minas Gerais. Stones are lighter than Colombian ones, mostly yellow-green, but they are often free of inclusions. Through deposits newly discovered since the beginning of the 1980s, Brazil has become one of the most important suppliers of emeralds.

Since the second half of the 1950s, emerald deposits have been exploited in Zimbabwe. Most important is the Sandawana mine in the south. Crystals are small, but of very good quality.

In the northern Transvaal (South Africa), emeralds are mined by modern methods using machinery (Cobra and Somerset Mines). Only five percent of the production is of good quality. Most stones are light or turbid and only suitable for cabochons.

Emerald deposits were discovered in 1830 in Russia in the Urals north of Yekaterinburg (Sverdlovsk), but the yield has fluctuated widely over the years. Good qualities are rare; most crystals are opaque, turbid, and slightly yellow-green. The host rock is a biotite mica shale, interfoliated with talc and chlorite.

Further emerald deposits are in Afghanistan, Australia (New South Wales, Western Australia), Ghana, India, Madagascar, Malawi, Mozambique, Namibia, Nigeria, Pakistan, Zambia, Tanzania, and the United States (North Carolina). The emerald mines of Cleopatra (perhaps worked as early as 2000 B.C.), east of Aswan in upper Egypt, are of historical interest only.

The Austrian deposits in the Habachtal Valley near Salzburg are well known. The host rock is biotite hornblende shale. The stones are of interest only to mineral collectors, as cuttable material is rare. The stones are mostly turbid; color quality is good. Some individual emeralds have been found in Norway, north of Oslo.

Famous Emeralds   There are many well-known large emeralds, as valuable and famous as diamonds and rubies. Some beautiful specimens of several hundred carats are kept by the British Museum of Natural History in London, by the American Museum of Natural History in New York, in the treasury of Russia, in the state treasury of Iran, and in the Treasury room in Topkapi Palace, Istanbul, Turkey. In the Viennese treasury is a vase, 4¾ in (12 cm) high and weighing 2205ct, cut from a single emerald crystal.

Working   Because emerald is so sensitive to knocks, a step cut was developed, the four corners being truncated by facets (so-called emerald cut). Clear, transparent qualities are sometimes brilliant cut. Turbid stones are used only for cabochons or as beads for necklaces. Occasionally emeralds are worn in their natural form, and sometimes engraved.

For some years, a number of cutters in Israel have been specializing in cutting emerald, after diamond cutters lost much of their production to India.

Possibilities for Confusion   Due to very similar colors, possibilities of confusion exist with aventurine (page 122), demantoid (page 106), diopside (page 190), dioptase (page 194), fluorite (page 198), grossularite (page 106), hiddenite (page 114), peridot (page 158), uvarovite (page 106), and verdelite (page 110).

Numerous doublets are on the market, mostly two genuine pale stones (rock crystal, aquamarine, beryl, or pale emerald) cemented together with emerald-green paste. The pavilion or both top and bottom may also be glass or synthetic spinel. The upper parts of natural stones are determined by inclusions and hardness, which are features of genuineness. When set, these doublets can be difficult to detect.

The first emerald synthesis was made in 1848. Since the turn of the century, various methods have been developed, and, since the 1950s, commercial products of excellent quality have appeared on the market. An important aid to differentiation is ultraviolet light. Synthetic emerald transmits ultraviolet light more than natural emerald.

In order to hide very fine hairline fractures and other faults, emeralds are often dipped into special oils or prepared with artificial resin in a vacuum; typically this is done at the country of origin. In the United States, FTC guidelines require disclosure of this treatment.

Aquamarine   Beryl Species

Color: Light blue to dark blue, blue-green	Transparency: Transparent to opaque
Color of streak: White	Refractive index: 1.564–1.596
Mohs' hardness: 7½–8	Double refraction: −0.004 to −0.005
Density: 2.68–2.74	Dispersion: 0.014 (0.009–0.013)
Cleavage: Indistinct	Pleochroism: Definite; nearly colorless, light
Fracture: Conchoidal, uneven, brittle	blue, blue, light green
Crystal system: (Hexagonal), hexagonal	Absorption: 537, 456, 427
prisms	Maxixe-A: 695, 655, 628, 615, 581, 550
Chemical composition: $Al_2Be_3Si_6O_{18}$	Fluorescence: None
aluminum beryllium silicate	

Aquamarine (Latin—water of the sea) is so named because of its seawater color. A dark blue is the most desired color. The coloring agent is iron. Lower qualities are heated to 725–850 degrees F (400–450 degrees C) to change them to the desired, permanent aquamarine blue. Higher heat will lead to discoloration. Care must be taken when making jewelry! Colors can also be improved with neutron and gamma irradiation, but these changes do not last. Aquamarine is brittle and sensitive to pressure. Inclusions of fine, oriented hollow rods or aligned foreign minerals rarely cause a cat's eye effect or asterism with six-rayed stars with a vivid sheen.

Santa Maria   Trade term (i.e., not an established variety name) for especially fine aquamarine. Named after the mine with the same name in Ceara (Brazil).

Santa-Maria-Africana   Trade term for fine aquamarine from Mozambique, on the market since 1991. Name derived from the "Santa Maria" quality of Ceara (Brazil).

Maxixe   Deep blue beryl; color fades in daylight. Originally (since 1917) it came only from the Maxixe Mine in Minas Gerais (Brazil). Since the 1970s it has been more widely available, but obviously made more beautiful through irradiation, but the color does not last.

Deposits   The most important deposits are in Brazil, spread throughout the country. The well-known deposits in Russia (the Urals) seem to be worked out. Other deposits of some commercial significance are in Australia (Queensland), Burma (Myanmar), China, India, Kenya, Madagascar, Mozambique, Namibia, Nigeria, Zambia, Zimbabwe, and the United States. The host rock is pegmatite and coarse-grained granite as well as their weathered material.

The largest aquamarine of gemstone quality was found in 1910 in Marambaya, Minas Gerais (Brazil). It weighed 243 lb (110.5 kg), was 18 in (48.5 cm) long and 15½ in (42 cm) in diameter, and was cut into many stones with a total weight of over 100,000 ct. There have been finds weighing a few tons, but these aquamarines are opaque and gray, not suitable for cutting.

The preferred cuts are step (emerald) and brilliant-cut with rectangular or long oval shapes. Turbid stones are cut en cabochon or are used for necklace beads.

Possibilities for Confusion   With euclase (page 178), kyanite (page 196), topaz (page 102), tourmaline (page 110), zircon (page 108), and glass imitations. Synthetic aquamarine can be produced but is uneconomical. The "synthetic aquamarine" sold in the trade is really aquamarine-colored synthetic spinel.

---

1 Aquamarine, emerald cut, 72.46ct	5 Aquamarine, antique cut, 18.98ct
2 Aquamarine, emerald cut, 17.41ct	6 Aquamarine, briolette, 6.65ct
3 Aquamarine, antique cut, 45.38ct	7 Aquamarine, cyrstal, 68.5mm, 45g
4 Aquamarine, marquise, 25.58ct	8 Aquamarine, 3 crystals, together 77g

1

2

3

4

5

6

7

8

Precious Beryl

Color: Gold-yellow, yellow-green, yellow, pink, colorless
Color of streak: White
Mohs' hardness: 7½–8
Density: 2.66–2.87
Cleavage: Indistinct
Fracture: Conchoidal, brittle
Crystal system: (Hexagonal), hexagonal prisms
Chemical composition: $Al_2Be_3Si_6O_{18}$ aluminum berylium silicate
Transparency: Transparent to opaque

Beryl Species

Refractive index: 1.562–1.602
Double refraction: –0.004 to –0.010
Dispersion: 0.014 (0.009–0.013)
Pleochroism: Golden: weak; lemon-yellow, yellow
Heliodor: weak; golden-yellow, green-yellow
Morganite: definite; pale pink, bluish-pink
Green: definite; yellow-green, blue-green
Absorption spectrum: Not diagnostic
Fluorescence: Morganite: weak; violet

Precious beryl refers to all color varieties of the beryl group that are not called emerald (page 90) and aquamarine (page 94).

Precious beryls are brittle and therefore sensitive to pressure and resistant to chemicals used in the household, and they have a vitreous luster, occasionally displaying the cat's-eye effect and asterism. They are typically found with aquamarine (page 94). Often used with a step cut. Color varieties have either special names in the trade, or the respective color precedes the word beryl (e.g., yellow beryl, green beryl).

Bixbite [3]   Raspberry-red. Origin of the name is unknown. Many scientists reject this as separate variety. Occurrences in Utah (United States).

Golden Beryl [1]   Color varies between lemon-yellow and golden-yellow. Inclusions are rare. Decolorization at 482 degrees F (250 degrees C). Deposits in Brazil, Madagascar, Namibia, Nigeria, Zimbabwe, Sri Lanka.

Goshenite [5]   Colorless beryl, named after locality in Goshen, Massachusetts (United States). Used as imitation for diamond and emerald by applying silver or green metal foil to the cut stone. Occurrences in Brazil, China, Canada, Mexico, Russia, and the United States.

Heliodor [2,6]   Light yellow-green (Greek—present of the sun). Discovered in 1910 as apparent new variety in Namibia, but beryls of the same color were previously known in Brazil and Madagascar. Since there is no clear distinction possible in the yellow and green-yellow tones in comparison to golden beryl, heliodors are generally rejected as an independent precious beryl variety and rather are counted among the weak-colored golden beryls.

Morganite [4]   (Also called pink beryl) Soft pink to violet, also salmon-colored. Inclusions are rare. Named after the American banker and collector J. P. Morgan. Density between 2.71 and 2.90. Inferior qualities can be improved by heating above 752 degrees F (400 degrees C). Deposits in Afghanistan, Brazil, China, Madagascar, Mozambique, Namibia, Zimbabwe, and the United States (Utah, California). Synthetic morganite is known.

Possibilities for Confusion   With many gemstones because of the richness of colors of the precious beryls. Greenish precious beryls are transferred into blue aquamarine stones through heating at 725–930 degrees F (444–500 degrees C).

The demarcation between emerald (page 90) and green beryl depends on the depth and intensity of color as well as the hue. To be considered emerald, the stone's color must be reasonably strong and green, bluish green, or perhaps slightly yellowish green. If the color is too pale or too yellowish, the stone is classified as green beryl.

1 Golden beryl, antique cut, 28.36ct
2 Heliodor, antique cut, 45.24ct
3 Bixbite, antique cut, 49.73ct
4 Morganite, antique cut, 23.94ct

5 Goshenite, marquise, 25.58ct
6 Heliodor, oval, 29.79ct
7 Beryl, 2 crystals, together 32.5g
8 Morganite, rough, 24.5g

Chrysoberyl

Color: Golden-yellow, green-yellow, green, brownish, red
Color of streak: White
Mohs' hardness: 8½
Density: 3.70–3.78
Cleavage: Good
Fracture: Weak conchoidal, uneven
Crystal system: (Orthorhombic), thick-tabled, intergrown triplets
Chemical composition: BeAl₂O₄ beryllium

aluminum oxide
Transparency: Transparent to opaque
Refractive Index: 1.746–1.763
Double refraction: +0.007 to +0.011
Dispersion: 0.015 (0.011)
Pleochroism: Very weak: red to yellow, yellow to light green, green
Absorption: 504, 495, 485, <u>445</u>
Fluorescence: Usually none
Green: weak; dark red

Chrysoberyl (Greek—gold) has been known since antiquity; the varieties alexandrite and chrysoberyl cat's-eye are especially valued. The host rock is granite pegmatite, mica schist, and placers. Deposits of the actual chrysoberyl (numbers 3, 9, and 10 below) are in Brazil (Minas Gerais, Espirito Santo) and Sri Lanka, as well as Burma (Myanmar), Madagascar, Russia (Urals), Zimbabwe, and the United States.

Stones are fashioned mainly in step, Ceylon, and brilliant cuts. The famous Hope chrysoberyl (London), a light green, faceted stone of 45ct is completely clean.

Possibilities for Confusion With andalusite (page 178), brasiliante (page 190), golden beryl (page 96), hiddenite (page 114), peridot (page 158), sapphire (page 86), sinhalite (page 186), skapolite (page 188), spinel (page 100), topaz (page 102), tourmaline (page 110), and zircon (page 108).

Chrysoberyl Cat's-Eye [2, 4] (Also called cymophane; Greek—waving light) Fine, parallel inclusions produce a silver-white line, which appears as a moving light ray in a cabochon cut stone. The name chrysoberyl cat's-eye is derived from this effect, which reminds one of the pupil of a cat. The short term "cat's-eye" always refers to chrysoberyl; all other cat's-eye must be designated by an additional name. There are deposits in Sri Lanka and Brazil, as well as in China, India, and Zimbabwe.

Possibilities for Confusion With quartz cat's-eye (page 124), prehnite cat's-eye (page 188). Synthetic chrysoberyl cat's-eye and doublets are known.

Alexandrite [5–8] (Named after Czar Alexander II) Was discovered only as recently as 1830 in the Urals. It is green in daylight, and light red in artificial incandescent light. This changing of color is best seen in thick stones. Alexandrite displaying the cat's-eye effect is a great rarity. Care must be taken when working with it, as it is sensitive to knocks and color changes are possible with exposure to great heat. High-quality alexandrite is one of the most expensive of all gemstones.

The deposits in the Urals are worked out. Today it is mined in Sri Lanka and Zimbabwe, and since the end of the 1980s especially in Brazil (Minas Gerais). Deposits are also found in Burma, Madagascar, and Tanzania. The largest stone, 1876ct, was found in Sri Lanka. The largest cut alexandrite weighs 66ct; it is in the Smithsonian Institution in Washington, D.C.

Possibilities for Confusion With synthetic corundum, synthetic spinel (page 243), andalusite (page 178), pyrope (page 104). Synthetic alexandrite is available on the market.

---

1 Chrysoberyl in host rock
2 Chrysoberyl cat's-eye, 24.09ct
3 Chrysoberyl, 3.36 and 2.23ct
4 Chrysoberyl cat's-eye, 4.33ct
5 Alexandrite in daylight and artificial light
6 Alexandrite, oval, 0.80ct

7 Alexandrite, intergrown triplet crystal
8 Alexandrite cat's-eye, 2.48ct
9 Chrysoberyl, oval, 9.24ct
10 Chrysoberyl, antique cut, 2.11ct
11 Chrysoberyl crystal
12 Chrysoberyl in host rock

Spinel

Color: Red, orange, yellow, brown, blue, violet, purple, green, black
Color of streak: White
Mohs' hardness: 8
Density: 3.54–3.63
Cleavage: Indistinct
Fracture: Conchoidal, uneven
Crystal system: (Cubic), octahedron, twins, rhombic dodecahedron
Chemical composition: MgAl$_2$O$_4$

magnesium aluminium oxide
Transparency: transparent to opaque
Refractive index: 1.712–1.762
Double refraction: None
Dispersion: 0.020 (0.011)
Pleochroism: Absent
Absorption: red Sp.: <u>685</u>, <u>684</u>, <u>675</u>, <u>665</u>, 656, 650, 642, 632, <u>595–490</u>, 465, 455
Fluorescence: Red Sp. strong: red; Blue Sp. weak: reddish, green

In mineralogy, spinel classifies a whole group of related minerals; only a few are gemstone quality. The origin of name is uncertain, perhaps Greek "spark" or Latin "thorn." The species gemologists designate as spinel occurs in all colors, the favorite being a ruby-like red. The coloring agents are iron, chromium, vanadium, and cobalt. Large stones are rare and star spinels are very rare.

Flame spinel   Trade term for bright orange to orange-red spinel. In the past, sometimes called rubicelle.

Balas spinel [4]   Falsely called Balas ruby (after a region of Afghanistan). Pale red variety.

Pleonaste [1]   Also called Ceylonite. Dark green to blackish, opaque spinel which contains iron (Mg,Fe) Al$_2$O$_4$; density 3.63–3.90.

Hercynite   Dark green to black spinel-group species, containing iron. FeAl$_2$O$_4$; density 3.95.

Gahnite   Also known as Zinc spinel. Blue, violet, or dark green to blackish spinel-group mineral. ZnAl$_2$O$_4$; density 4.00–4.62. (See page 204, no. 1).

Gahnospinel [3,7]   Blue to dark blue or green species between spinel and gahnite in composition, colored by iron. (MgZn)Al$_2$O$_4$; density 3.58-4.06.

Picotite   Also called Chrome spinel. Brownish, dark green, or blackish spinel-group member. Fe (Al,Cr$_2$)O$_4$; density 4.42.

Spinel was recognized as an individual mineral only 150 years ago. Before then it was classed as ruby. Some well-known "rubies" are really spinels, such as the "Black Prince's Ruby" in the English Crown (see photo, page 9) and the 361ct "Timur Ruby" in a necklace in the English Crown Jewels. The drop-shaped spinels in the Wittelsbach crown of 1830 were also originally thought to be rubies.

It occurs with ruby and sapphire in placer deposits, mainly in Burma (Myanmar, near Mogok), Cambodia, and Sri Lanka (near Ratnapura). Other deposits are found in Afghanistan, Australia, Brazil, Madagascar, Nepal, Nigeria, Tadzhikistan, Tanzania, Thailand, and the United States (New Jersey).

The two largest spinels (formed as roundish octahedrons) weigh 520ct each and are in the British Museum of London. The Diamond Fund in Moscow owns a spinel that weighs over 400ct.

Possibilities for Confusion   With amethyst (page 118), chrysoberyl (page 98), pyrope (page 104), sapphire (page 86), topaz (page 102), tourmaline (page 110), and zircon (page 108). Synthetic spinels have been on the market since the 1920s (compare to page 246). They imitate not only natural spinel but also many other gemstones, especially ruby.

---

1 Pleonaste crystals in host rock
2 Spinel, 28.47 and 4.16ct
3 Spinel, 3 faceted stones
4 Spinel, so-called balas ruby 17.13ct
5 Spinel, antique cut, 5.05ct
6 Spinel, two ovals, 7.96 and 5.32ct

7 Spinel, blue, 15.08 and 30.11ct
8 Spinel, 12 different reds
9 Spinel, yellow, 3.14 and 5.07ct
10 Spinel, crystals and other rough stones

Topaz

Color: Colorless, yellow, orange,
  red-brown, light to dark blue,
  pink-red, red, violet, light green
Color of streak: White
Mohs' hardness: 8
Density: 3.49–3.57
Cleavage: Perfect
Fracture: Conchoidal, uneven
Crystal system: Orthorhombic, prisms with
  multi-faceted ends, often 8-sided in
  cross-section striations along length
Chemical composition: $Al_2SiO_4(F,OH)_2$
  fluor containing aluminium silicate

Transparency: Transparent, translucent
Refractive index: 1.609–1.643
Double refraction: +0.008 to +0.016
Dispersion: 0.014 (0.008)
Pleochroism: Yellow: definite; lemon-honey,
  straw-yellow
  Blue: weak; light and dark blue,
  Red: strong; dark red, yellow, pink-red
Absorption spectrum: Pink: <u>682</u>
Fluorescence: Pink: weak; brown
  Red: weak; yellow-brown
  Yellow: weak; orange-yellow

Formerly, the name topaz was not applied consistently or specifically; one called all yellow and golden-brown, and sometimes also green, gemstones topaz. The name topaz is most probably derived from an island in the Red Sea, now Zabargad but formerly Topazos, the ancient source of peridot (see page 158).

Colors of the gemstone that is today called topaz are rarely vivid. The most common color is yellow with a red tint; the most valuable is pink to reddish-orange. The coloring agents are iron and chromium. Some yellowish-brown varieties of certain deposits gradually fade in the sunlight.

Care must be taken during polishing and setting because of the danger of cleavage. They are also not resistant to hot sulphuric acid. The luster is vitreous.

Deposits are associated with pegmatites or secondary placers. During the 18th century, the most famous topaz mine was at Schneckenstein in the southern Voigtland in Saxony. Today, Brazil (Minas Gerais) is the most important supplier. Other deposits are in Afghanistan, Australia, Burma (Myanmar), China, Japan, Madagascar, Mexico, Namibia, Nigeria, Pakistan, Russia (the Urals, Transbaikalia), Zimbabwe, Sri Lanka, and the United States. Light blue topazes are found also in Northern Ireland, Scotland, and Cornwall, England.

Topazes weighing several pounds are known. In 1964 some blue topazes were found in the Ukraine, each weighing about 220 lb (100 kg). The Smithsonian Institution in Washington, D.C., owns cut topazes of several thousand carats each.

Colored stones are usually step- (emerald) or scissor-cut, and colorless ones or weakly colored ones are brilliant-cut. Topazes with disordered inclusions are cut en cabochon.

Possibilities for Confusion With apatite (page 194), aquamarine (page 94), brazilianite (page 190), chrysoberyl (page 98), citrine (page 120), danburite (page 182), diamond (page 70), precious beryl (page 96), fluorite (page 198), kunzite (page 114), orthoclase (page 164), phenakite (page 180), ruby (page 82), sapphire (page 86), spinel (page 100), tourmaline (page 110), and zircon (page 108).

Since 1976, blue synthetic topazes are known. Almost all blue topaz sold today is produced by first irradiating and then heating natural colorless topaz. Since the quartz variety citrine (page 120) is in the trade often falsely called "gold topaz" or "Madeira topaz," real topaz is sometimes called precious topaz, in order to clearly distinguish them.

---

1 Topaz, rectangular step cut,
  46.61ct, Brazil
2 Topaz, rough, 225.50ct, Brazil
3 Topaz, crystal cleavage, 18.00 ct,
  Brazil
4 Topaz, oval, 93.05ct, Afghanistan
5 Topaz, emerald cut, 88.30ct, Brazil

6 Topaz, fancy cut, 32.44ct
7 Topaz, 2 faceted ovals, 53.75ct,
  Russia
8 Topaz, cabochon, 17.37ct, Brazil
9 Topaz, crystal, 65.00ct, Brazil
10 Topaz, crystal in host rock

Garnet Group

This is a group of differently colored minerals with similar crystal structure and related chemical composition. The main representatives are pyrope, almandite and spessartite (pyralspite series), grossularite, andradite, and uvarovite (ugrandite series). Within the series are also mixed members. The name derives from the Latin for grain because of the rounded crystals and similarity to the red kernals of the pomegranate. Garnet, in the popular sense, is usually understood only as the red "carbuncle stones" pyrope and almandite.

Data common to all garnets:

Color or streak: White	Fracture: Conchoidal, splintery, brittle
Mohs' hardness: 6½–7½	Transparency: Transparent to opaque
Cleavage: Indistinct	Double refraction: Normally none
Crystal system: (Cubic) rhombic dodecahedron, icositetrahedron	Pleochroism: Absent
	Fluorescence: Mostly none
	Luster: Vitreous

Pyrope [4, 5]

Color: Red, frequently with brown tint
Density: 3.62–3.87
Chemical composition: $Mg_3Al_2(SiO_4)_3$ magnesium aluminium silicate

Refractive index: 1.720–1.756
Dispersion: 0.022 (0.013–0.016)
Absorption: <u>687</u>, <u>685</u>, 671, 650, <u>620–520</u>, 505

Pyrope (Greek—fiery) was the fashion stone of the 18th and 19th centuries, especially the "Bohemian Garnet." Deposits are found in Burma (Myanmar), China, Madagascar, Sri Lanka, South Africa, Tanzania, and the United States. Can be confused with almandite (see below), ruby (page 82), spinel (page 100), and tourmaline (page 110). Imitations are made with red glass.

Rhodolite [6, 7, 8]   Purplish red or rose-color garnet between pyrope and almandite in composition.

Almandite [9,10]   Garnet Group

Color: Red with violet tint
Density: 3.93–4.30
Chemical composition: $Fe_3Al_2(SiO_4)_3$ iron aluminium silicate

Refractive index: 1.770–1.820
Dispersion: 0.027 (0.013–0.016)
Absorption: 617, <u>576</u>, <u>526</u>, <u>505</u>, 476, 462, 438, 428, 404, 393

Its name is derived from the town in Asia Minor. Deposits are found in Brazil, India, Madagascar, Sri Lanka, and the United States, as well as the Czech Republic and Austria. Can be confused with pyrope (see above), ruby (page 82), spinel (page 100), and tourmaline (page 110).

Spessartite [2, 3]   Garnet Group

Color: Orange to red-brown
Density: 4.12–4.18
Chemical composition: $Mn_3Al_2(SiO_4)_3$ manganese aluminum silicate

Transparency: Transparent, translucent
Refractive index: 1.790–1.820
Dispersion: 0.027 (0.015)
Absorption: 495, 485, 462, <u>432</u>, 424, <u>412</u>

Its name is derived from occurrence in the Spessart (= forest), Germany. Deposits are found in Burma (Myanmar), Brazil, China, Kenya, Madagascar, Sri Lanka, Tanzania, and the United States.

The best specimens come from Namibia ("Mandarin Spessartite"). Can be confused with andalusite (page 178), chrysoberyl (page 98), fire opal (page 152), hessonite (page 106), sphene (page 194), and topaz (page 102).

---

1 Range of garnet colors, green-yellow-brown-red
2 Spessartite crystal in host-rock
3 Spessartite, 3 cabochons
4 Pyrope crystal, icosatetrahedron
5 Pyrope, 3 faceted stones

6 Rhodolite, brilliant cut, 4.02ct
7 Rhodolite, marquise, 2 ovals
8 Rhodolite crystal, rolled
9 Almandite in mica
10 Almandite, trapezoid, oval, emerald cut

Grossularite (Grossular) [4, 5]
Garnet Group

Color: Colorless, green, yellow, brown
Density: 3.57–3.73
Chemical composition: Ca₃Al₂(SiO₄)₃
calcium aluminum silicate

Refractive index: 1.734–1.759
Dispersion: 0.020 (0.012)
Absorption spectrum: 697, 630, 605, 505
Fluorescence: Dense G. Strong: red-orange

Grossularite (Latin—gooseberry) deposits are found in Canada, Kenya, Mali, Pakistan, Russia (Siberia), Sri Lanka, South Africa, Tanzania, and Vermont. Can be confused with demantoid (page 106), emerald (page 90), and tourmaline (page 110).

Hessonite [1, 2, 3] (also called cinnamon stone and Kaneel stone) Brown-red variety. Deposits are found in Sri Lanka, as well as Brazil, India, Canada, Madagascar, Tanzania, and the United States. Can be confused with chrysoberyl (page 98), cassiterite (page 184), spessartite (page 104), and zircon (page 108).

Leuco garnet [6] Colorless variety. Deposits are found in Canada, Mexico, and Tanzania.

Hydrogrossular (also falsely called Transvaal jade and garnet jade) Dense, opaque greenish variety. Deposits are in South Africa, Burma (Myanmar), and Zambia.

Tsavorite (Tsavolite) Green to emerald green variety from Kenya and Tanzania; discovered in the early 1970s.

Andradite
Garnet Group

Color: Black, brown, yellow-brown
Density: 3.7–4.1
Chemical composition: Ca₃Fe₂(SiO₄)₃
calcium iron silicate

Refractive index: 1.88–1.94
Dispersion: 0.057
Absorption: 701, 693, 640, 622, 443

Andradite (named after a Portuguese mineralogist) varieties include the following:

Demantoid [7] the most valuable garnet (name means "diamond-like luster"), green to emerald green. Deposits found in China, Korea, Russia, the United States, and Zaire. Can be confused with grossularite (see above), peridot (page 158), emerald (page 90), spinel (page 100), tourmaline (page 110), and uvarovite (see below).

Melanite [9] Opaque black variety (Greek—black). Deposits are in Germany (Kaiserstuhl/Baden-Württemberg), France, Italy, and Colorado. Used for mourning jewelry. Can be confused with black glass.

Topazolite [10] Yellow to lemon yellow, topaz-like (therefore the name) variety. Deposits found in Switzerland, the Italian Alps, and California.

Uvarovite [8]
Garnet Group

Color: Emerald-green
Density: 3.77
Chemical composition: Ca₃Cr₂(SiO₄)₃
calcium chromium silicate

Refractive index: 1.87
Dispersion: (0.014-0.021)
Absorption: Cannot be evaluated

Named after a Russian statesman. Rarely occurs in gemstone quality. Deposits found in Finland, India, Canada, Poland, Russia (the Urals), and California. Can be confused with demantoid (page 106) and emerald (page 90).

---

1 Hessonite crystal in host rock
2 Hessonite, 2 cabochons
3 Hessonite, 3 faceted stones
4 Grossularite, rhombic dodecahedron
5 Grossularite, green and copper-brown
6 Leuco garnet, marquise, 1.97ct

7 Demantoid, 3 rough and 3 cut stones
8 Uvarovite crystals, partly rolled
9 Melanite, 2 crystals
10 Topazolite, rough and faceted

Zircon

Color: Colorless, yellow, brown, orange, red, violet, blue, green
Color of streak: White
Mohs' hardness: 6½–7½
Density: 3.93–4.73
Cleavage: Indistinct
Fracture: Conchoidal, very brittle
Crystal system: (Tetragonal) short, stocky four-sided prisms with pyramidal ends
Chemical composition: $ZrSiO_4$ zirconium silicate
Transparency: Transparent to translucent
Refractive index: 1.810–2.024

Double refraction: +0.002 to +0.059
Low-z: None
Dispersion: 0.039 (0.022), Low z.: None
Pleochroism: Yellow z. very weak: honey-yellow, brown-yellow;
Blue z. distinct: blue, yellow-gray to colorless
Absorption: (High z.) 691, 689, 662, 660, 653, 621, 615, 589, 562, 537, 516, 484, 460, 433
Fluorescence: Blue: very weak; light orange Red and brown: weak; dark yellow

Zircon has been known since antiquity, albeit under various names. Today's name is most likely derived from the Persian language ("golden colored"). Because of its high refractive index and strong dispersion, it has great brilliance and intensive fire. It is brittle and therefore sensitive to knocks and pressure; the edges are easily damaged. The luster is vitreous to a brilliant sheen. A content of radioactive elements (uranium, thorium) causes large variations of physical properties. Zircons with the highest values in optical properties are designated as high zircons, those with the lowest values as low zircons. In between are the medium zircons. The alteration caused by radioactive elements in green (low) zircons is so advanced that these stones can be nearly amorphous, even though their outer appearance seems unchanged. These green, slightly radioactive zircons are rarely found in the gemstone trade, but are highly prized by collectors. Zircons with a cat's-eye effect are also known.

Hyacinth   Old term for yellow, yellow-red to red-brown zircon. Also used for hessonite (page 106).

Jargon   Old term for straw-yellow to almost colorless zircon.

Starlight   Trade term for blue zircon variety, created through heating other zircons.
Deposits are mainly alluvial; found in Burma (Myanmar), Cambodia, Sri Lanka, Thailand, as well as Australia, Brazil, Korea, Madagascar, Mozambique, Nigeria, Tanzania, and Vietnam.
In nature the gray-brown and red-brown zircons are the most common. Colorless specimens are rare. In the South Asian countries where found, the brown varieties are heat-treated at temperatures of 1472–1832 degrees F (800–1000 degrees C), producing colorless and blue zircons. These colors do not necessarily remain constant; ultraviolet rays or sunlight can produce changes. Colorless stones are brilliant cut; colored ones are given a brilliant or step (emerald) cut. Synthetic zircons are only of scientific interest.

Possibilities for Confusion   With aquamarine (page 94), chrysoberyl (page 98), demantoid (page 106), diamond (page 70), hessonite (page 106), idocrase (page 186), sapphire (page 86), sinhalite (page 186), topaz (page 102), and tourmaline (page 110). Colorless heat-treated zircon is fraudulently offered for diamond as "matura" ("matara") diamond.

---

1 Zircon, rectangular, 9.81ct
2 Zircon, pear and brilliant
3 Zircon, brilliant, 14.35ct
4 Zircon, 2 brilliants
5 Zircon, oval, 5.11ct
The photos are 20 percent larger than the originals.

6 Zircon, emerald cut, 7.92ct
7 Zircon, emerald cut, 4.02ct
8 Zircon, 4 brilliants
9 Zircon, 3 faceted stones
10 Zircon, rough stones

Tourmaline Group

The tourmaline group refers to a number of related species and varieties.

Color: Colorless, pink, red, yellow, brown, green, blue, violet, black, multicolored
Color of streak: White
Mohs' hardness: 7–7½
Density: 2.82–3.32
Cleavage: Indistinct
Fracture: Uneven, small conchoidal, brittle
Crystal system: (Trigonal), long crystals with triangular cross section and rounded sides, definite striation parallel to main axis
Transparency: Transparent to opaque

Refractive index: 1.614–1.666
Double refraction: –0.014 to –0.032
Disperson: 0.017 (0.009–0.011)
Pleochroism: Red t.: definite: dark red, light red;
Brown t.: definite: dark brown, light brown;
Green t.: strong: dark green, yellow-green;
Blue t.: strong: dark blue, light blue
Absorption spectrum: Often extremely faint
Fluorescence: Weak or none

Even though tourmaline has been known since antiquity in the Mediterranean region, the Dutch imported it only in 1703 from Sri Lanka to Western and Central Europe. They gave the new gems a Sinhalese name, *Turamali*, which is thought to mean "stone with mixed colors."

According to color, the following varieties are recognized in the trade:

Achroite   (Greek—without color) colorless or almost colorless, quite rare.

Dravite   [page 113, nos. 1, 7, 8] yellow-brown to dark brown, sometimes used for stones not of the dravite species (see below).

Indicolite (indigolite)   [page 113, nos. 3, 5, 11, 15] named (after color) blue in all shades.

Rubellite   [page 113, nos. 2, 4] (Latin—reddish) pink to red, sometimes with a violet tint; ruby color is the most valuable.

Schorl [4, 5]   black, very common; used for mourning jewelry. Name derived from an old mining term, sometimes applied to tourmaline that is not actually schorl (see below).

Siberite   (after finds in Urals) lilac to violet blue.

Verdelite   [page 112, nos. 6, 13] ("green stone") green in all shades.

Recently, instead of variety names, more and more frequently color names are simply added to the word tourmaline, e.g., yellow tourmaline, pink tourmaline.

Mineralogy distinguishes tourmalines according to their chemical composition. The individual members include the following:

Buergerite   (after U.S. scholar) $NaFe_3Al_6(BO_3)_3Si_6O_{18}(O,F)_4$ = iron tourmaline

Dravite   (after a deposit near the river Drave, Carinthia/Austria) $NaMg_3Al_6(BO_3)_3Si_6O_{18}(OH)_4$ = magnesium tourmaline

Elbaite (after the island of Elba/Italy) $Na(Li,Al)_3Al_6(BO_3)_3Si_6O_{18}(OH)_4$ = lithium tourmaline

Liddicoatite   (after U.S. gemologist) $Ca(Li,Al)_3Al_6(BO_3)_3Si_6O_{18}(O,OH,F)_4$ = calcium tourmaline

Schörl   (after an old mining expression for "false ore") $NaFe_3Al_6(BO_3)_3Si_6O_{18}(OH)_4$ = iron tourmaline

Tsilaisite   (after a local name in Madagascar) $NaMn_3Al_6(BO_3)_3Si_6O_{18}(OH)_4$ = manganese tourmaline

Uvite   (after a province in Sri Lanka) $(Ca,Na)(Mg,Fe)_3Al_5Mg(BO_3)_3Si_6O_{18}(OH,F)_4$ = magnesium tourmaline

---

1 Tourmaline, 8 polished cross sections
2 Rubellite cat's-eye, 1.87ct
3 Tourmaline, crystals stem-like on quartz

4 Schorl, crystals, opaque
5 Schorl in quartz, partly polished
6 Tourmaline "watermelon"
7 Verdelite, 2 crystals
8 Multicolored tourmaline, crystal

Unicolored tourmaline crystals are quite rare. Most show various tones in the same crystal or even different colors [page 111, nos. 6, 8]. There are colorless tourmalines with black crystal ends, green ones with red crystal ends, and some with different-colored layers. There are stones whose core is red, the inner layer white, and the outer layer green—called "watermelon" [page 111, no. 6].

Tourmaline cat's-eyes exits in various colors, but only in the green and pink varieties [page 111, no. 2] is the chatoyancy usually strong, caused by thin tube-like inclusions. Some tourmalines show a slight change of color in artificial light. They have a vitreous sheen on crystal surfaces, a greasy sheen on fractured surfaces.

By heating and subsequent cooling, as well as by applying pressure, i.e., by rubbing, a tourmaline crystal will become electrically charged. It will then attract dust particles as well as small pieces of paper (pyro- and piezo-electricity). The Dutch, who first imported tourmaline into Europe, knew of this effect. They used a heated stone to pull ash out of their meerschaum pipes and thus called this strange stone *aschentrekker* (ash puller). For a long time this was the popular name for a tourmaline. Due to this pyroelectrical effect, tourmaline has to be cleaned more often than other gemstones.

Deposits are found in pegmatites and alluvial deposits. The most important tourmaline supplier is Brazil (Minas Gerais, Paraiba). Other deposits are in Afghanistan, Australia, Burma (Myanmar), India, Madagascar, Malawi, Mozambique, Namibia, Nepal, Nigeria, Pakistan, Russia, Zambia, Zimbabwe, Sri Lanka, Tanzania, the United States (California, Maine), and Zaire. In Europe, there are tourmaline deposits on Elba (Italy) and in Switzerland (Tessin).

The most desired colors are intense pink and green. It is used in different cuts. Because of the strong pleochrism, dark stones must be cut so that the table lies parallel to the main axis. In the case of pale stones, the table should be perpendicular to the long axis in order to obtain the deeper color.

By heating to 842–1202 degrees F (450–650 degrees C), color changes can be produced in some tourmalines. Some green tourmaline becomes emerald green; others are lightened. The color of tourmalines which have been changed with gamma-irradiation may fade. Synthetic tourmalines are used only for research purposes. The stones, offered as synthetic tourmaline, are really tourmaline-colored synthetic spinels.

Possibilities for Confusion   With many gemstones, due to the large variety of colors, especially amethyst (page 118), andalusite (page 178), chrysoberyl (page 98), citrine (page 120), demantoid (page 106), emerald (page 90), hiddenite (page 114), idocrase (page 186), kunzite (page 114), morganite (page 96), peridot (page 158), prasiolite (page 120), ruby (page 82), topaz (page 102), zircon (page 108), and glass imitations.

1 Dravite, 2 faceted stones
2 Rubellite, oval, 1.73ct
3 Indicolite, emerald cut, 6.98ct
4 Rubellite, 2 faceted stones, together 4.55ct
5 Indicolite, emerald cut and antique cut
6 Verdelite, oval, 19.88ct
7 Dravite, cabochon, 19.97ct
8 Dravite, 3 faceted stones
9 Rubellite, oval, 6.16ct

10 Indicolite, crystal
11 Indicolite, faceted, 2.97ct
12 Indicolite crystal
13 Verdelite, 2 cabochons, 9.77ct
14 Tourmaline, 2 faceted yellow-green stones
15 Indicolite, 3 cabochons
16 Tourmaline, multicolored, 24ct
17 Rubellite, 3 cabochons
18 Tourmaline, 3 crystals

The illustrations are 20 percent larger than the originals.

Spodumene Species

The name refers to the mineral spodumene (Greek—ash-colored) because the common non-gem crystals are mostly opaque, white to yellowish. For a long time gem varieties have been known as hiddenite and kunzite; since the 1970s some isolated transparent colorless varieties have been found. Most recently light yellow and green varieties have also been known. Rarely displays the cat's-eye effect.

Hiddenite [1–3, 8]

Spodumene Species

Color: Yellow-green, green-yellow, emerald-green
Color of streak: White
Mohs' hardness: 6½–7
Density: 3.15–3.21
Cleavage: Perfect
Fracture: Uneven, brittle
Crystal system: Monoclinic; prismatic, tabular
Chemical composition: $LiAlSi_2O_6$ lithium
aluminum silicate
Transparency: Transparent
Refractive index: 1.660–1.681
Double refraction: +0.014 to +0.016
Dispersion: 0.017 (0.010)
Pleochroism: Definite; blue-green, emerald-green, yellow-green
Absorption spectrum: <u>690</u>, <u>686</u>, 669, 646, <u>620</u>, <u>437</u>, 433
Fluorescence: Very weak; red-yellow

Named after A. E. Hidden who discovered this stone in 1879 in North Carolina. The coloring agent is chromium. Colors can gradually fade; has a strong vitreous luster. Usually used with the step cut (emerald). In order to display strong colors (due to pleochroism), the table facet must be perpendicular to the main axis of the stone. Deposits occur in granite pegmatite. Deposits found in Burma (Myanmar), Brazil (Minas Gerais), Madagascar, North Carolina, and California. Can be confused with chrysoberyl (page 98), diopside (page 190), emerald (page 90), peridot (page 158), precious beryl (page 96), and verdelite (page 110).

Kunzite [4–7]

Spodumene Species

Color: Pink-violet, light violet
Color of streak: White
Mohs' hardness: 6½–7
Density: 3.15–3.21
Cleavage: Perfect
Fracture: Uneven, brittle
Crystal system: Monoclinic; prismatic, tabular
Chemical composition: $LiAlSi_2O_6$ lithium
aluminum silicate
Transparency: Transparent
Refractive index: 1.660–1.681
Double refraction: +0.014 to +0.016
Dispersion: 0.017 (0.010)
Pleochroism: Definite; amethyst color, pale red, colorless
Absorption spectrum: Not diagnostic
Fluorescence: Strong; yellow-red, orange

Named after the U.S. mineralogist G. F. Kunz, who first described this gem in 1902. The coloring agent is manganese. Stones are mostly light colored; colors can fade. Brownish and green-violet types can be improved in color by heating to about 300 degrees F (150 degrees C). There are frequently aligned inclusions such as tubes or fractures. The stone has a vivid vitreous luster. The table facet must be perpendicular to the main axis of the stone. Deposits occur in granite pegmatite. The main producer is Brazil (Minas Gerais). Other deposits are found in Afghanistan, Burma (Myanmar), Madagascar, Pakistan, and the United States. Can be confused with many pink-colored stones, especially amethyst (page 118), morganite (page 96), petalite (page 188), rose quartz (page 122), rubellite (page 110), sapphire (page 86), and topaz (page 102), as well as colored glass.

1 Hiddenite, emerald cut, 22.03ct	5 Kunzite, oval, 3.13ct
2 Hiddenite, pear-shaped, 9.30ct	6 Kunzite, antique cut, 6.11ct
3 Hiddenite, emerald cut, 19.14ct	7 Kunzite, 2 crystals
4 Kunzite, emerald cut, 16.32ct	8 Hiddenite, crystal and broken piece

Quartz

Quartz (named after a Slavic word for "hard") is the name for a group of minerals of the same chemical composition (SiO_2) and similar physical properties.

Macrocrystalline quartz   (crystals recognizable with the naked eye) includes stones gemologists classify as varieties of the quartz species: amethyst, aventurine, rock crystal, blue quartz, citrine, hawk's-eye, prasiolite, quartz cat's-eye, smoky quartz, rose quartz, and tiger's-eye.

Cryptocrystalline quartz   (microscopically small crystals) generally known as chalcedony, includes agate, petrified wood, chrysoprase, bloodstone, jasper, carnelian, moss agate, and sard.

Rock Crystal [8–11]   Quartz Species

Color: Colorless	Transparency: Transparent
Color of streak: White	Refractive index: 1.544–1.553
Mohs' hardness: 7	Double refraction: +0.009
Density: 2.65	Dispersion: 0.013 (0.008)
Cleavage: None	Pleochroism: Absent
Fracture: Conchoidal, very brittle	Absorption spectrum: None
Crystal system: (Trigonal), hexagonal prisms	Fluorescence: None
Chemical composition: SiO_2 silicon dioxide	

The name crystal comes from the Greek for "ice," as it was believed that rock crystal was eternally frozen. Rock crystals weighing many tons have been found. Cuttable material is rare. Inclusions are of goethite [star quartz, no. 12], gold, pyrite, rutile, and tourmaline [page 111, no. 5]; the luster is vitreous. Important deposits, among others all over the world, are found in Brazil, Madagascar, the United States, and the Alps. They are used for costume jewelry and delicate bowls and to imitate diamonds. Rock crystal alters to a smoky color with gamma irradiation. Can be confused with many colorless gems as well as glass. Synthetic rock crystal is used only for industrial purposes.

Smoky Quartz [1–7]   Quartz Species

Color: Brown to black, smoky gray	Transparency: Transparent
Color of streak: White	Refractive index: 1.544–1.553
Mohs' hardness: 7	Double refraction: +0.009
Density: 2.65	Dispersion: 0.013 (0.008)
Cleavage: None	Pleochroism: Dark: definite; brown, reddish-brown
Fracture: Conchoidal, very brittle	
Crystal system: (Trigonal), hexagonal prisms	Absorption spectrum: Cannot be evaluated
Chemical composition: SiO_2 silicon dioxide	Fluorescence: Usually none

Named after its smoky color. Very dark stones are called "morion" and "caingorm" [6,7]. Coloring is caused by (natural and artificial) gamma rays. The name *smoky topaz* is improper and no longer acceptable in the trade. Frequent inclusions are rutile needles [nos. 1, 2]. Deposits in Brazil, Madagascar, Russia, Scotland, Switzerland, and Ukraine. It is used just as rock crystal. Can be confused with andalusite (page 178), axinite (page 182), idocrase (page 186), sanidine (page 204), and tourmaline (page 110).

---

1 Smoky quartz with rutile inclusions	8 Rock crystal, 4 stones, faceted and cabochon
2 Smoky quartz with rutile, cabochon	
3 Smoky quartz, oval, 3.8g	9 Rock crystal, crystals and twins
4 Smoky quartz, 2 crystals	10 Rock crystal, brilliant cut, 5g
5 Smoky quartz, emerald cut, 5.6g	11 Rock crystal, baguette, 1.8g
6 Smoky quartz, oval, 6.2g	12 Star quartz, 15g
7 Smoky quartz crystal	

Amethyst [4–8]

Color: Purple, violet, pale red-violet	Transparency: Transparent
Color of streak: White	Refractive index: 1.544–1.553
Mohs' hardness: 7	Double refraction: +0.009
Density: 2.65	Dispersion: 0.013 (0.008)
Cleavage: None	Pleochroism: Weak: reddish-violet, gray-violet
Fracture: Conchoidal, very brittle	
Crystal system: (Trigonal), hexagonal prisms	Absorption spectrum: (550–520)
Chemical composition: SiO$_2$ silicon dioxide	Fluorescence: Weak; bluish

Amethyst is the most highly valued stone in the quartz group. The name means "not drunken" (Greek), as amethyst was worn as an amulet against drunkenness. Crystals are always grown onto a base. Prisms are usually not well developed, therefore are often found as crystal points (pointy amethyst) with the deepest color. These parts are broken off at the base for further treatment.

Heat treatment between 878–1382 degrees F (470–750 degrees C) produces light yellow, red-brown, green, or colorless varieties. There are some amethysts that lose some color in daylight. The original color can be restored by X-ray radiation. The coloring agent is iron. In artificial light, amethyst does not display as desirable qualities.

Found in geodes in alluvial deposits. The most important deposits are in Brazil ("Palmeira" amethyst of Rio Grande do Sul, "Maraba" amethyst of Para), Madagascar, Zambia, Uruguay, as well as in Burma (Myanmar), India, Canada, Mexico, Namibia, Russia, Sri Lanka, and the United States (Arizona). The best stones are faceted; others are tumbled or worked into ornaments. Formerly amethyst was a favorite gemstone of high officials of the Christian church. Can be confused with precious beryl (page 96), fluorite (page 198), kunzite (page 114), spinel (page 100), topaz (page 102), tourmaline (page 110), and tinted glass. Synthetic amethyst is abundant on the gemstone market.

Ametrine (also called trystine)   Color-zoned quartz variety, which consists half of amethyst and other half of citrine. Deposits are found in Brazil (Rio Grande do Sul) and Bolivia.

Amethyst Quartz [1–3]

Color: Violet with whitish stripes	Transparency: Translucent, opaque
Color of streak: White	Refractive index: 1.54–1.55
Mohs' hardness: 7	Double refraction: +0.009
Density: 2.65	Dispersion: 0.013 (0.008)
Cleavage: None	Pleochroism: Absent
Fracture: Conchoidal, brittle	Absorption spectrum: Cannot be evaluated
Crystal system: (Trigonal) compact	Fluorescence: None
Chemical composition: SiO$_2$ silicon dioxide	

Amethyst quartz is the rougher, more compact formation of amethyst, layered and striped with milky quartz. Occurs together with amethyst. It is used for beads, baroque stones, cabochons, and ornamental objects. Can be confused with striped fluorite (page 198).

---

1 Amethyst quartz, rough	5 Amethyst, 4 faceted stones
2 Amethyst quartz, 7 cabochons	6 Amethyst, double-ended crystal
3 Amethyst quartz, polished slice	7 Amethyst, brilliant cut, 4.16ct
4 Amethyst, marquise, 3.94ct	8 Amethyst, geode on agate

Citrine [1–6]

Color: Light yellow to dark yellow, gold-brown
Color of streak: White
Mohs' hardness: 7
Density: 2.65
Cleavage: None
Fracture: Conchoidal, very brittle
Crystal system: Hexagonal (trigonal); hexagonal prisms with pyramids
Chemical composition: SiO_2 silicon dioxide

Transparency: Transparent
Refractive index: 1.544–1.553
Double refraction: +0.009
Dispersion: 0.013 (0.008)
Pleochroism: Natural: weak; yellow-light yellow
Heat-treated: none
Absorption spectrum: Not diagnostic
Fluorescence: None

The name is derived from its lemon yellow color. The coloring agent is iron. Natural citrines are rare. Most commercial citrines are heat-treated amethysts (page 118) or smoky quartzes (page 116). Brazilian amethyst turns light yellow at 878 degrees F (470 degrees C) and dark yellow to red-brown at 1022–1040 degrees F (550–560 degrees C). Some smoky quartzes turn into citrine color already at about 390 degrees F (200 degrees C).

Almost all heat-treated citrines have a reddish tint. The natural citrines are mostly pale yellow. Names for citrine such as Bahia, Madeira, or Rio Grande topaz are improper and no longer accepted in the trade as they are deceptive. On the other hand, when one, for example, speaks of Madeira color and/or Madeira citrine, this is a correct usage; the expert properly connects a certain color with the locality name.

Deposits of natural-colored citrines are found in Brazil, Madagascar, and the United States, as well as in Argentina, Burma (Myanmar), Namibia, Russia, Scotland, and Spain. Well-colored citrines are used as ring stones and pendants; less attractive stones are made into necklaces or ornaments.

Can be confused with many yellow gemstones, especially apatite (page 194), golden beryl (page 96), orthoclase (page 164), topaz (page 102), and tourmaline (page 110), as well as tinted glass.

Prasiolite [7, 8]

Color: Leek-green
Color of streak: White
Mohs' hardness: 7
Density: 2.65
Cleavage: None
Fracture: Conchoidal, very brittle
Crystal system: Hexagonal (trigonal); hexagonal prisms
Chemical composition: SiO_2 silicon dioxide

Transparency: Transparent
Refractive index: 1.544–1.553
Double refraction: +0.009
Dispersion: 0.013 (0.008)
Pleochroism: Very weak; light green, pale green
Absorption spectrum: Not diagnostic
Fluorescence: None

Prasiolite (Greek—leek-green stone) is not found in nature. It is produced by heating violet amethyst or yellowish quartz from the deposit Montezuma in Minas Gerais, Brazil, to a temperature of about 930 degrees F (500 degrees C). Other deposits of heatable amethyst have recently been reported in Arizona. In sunlight, the color commonly fades.

Can be confused with precious beryl (page 96), peridot (page 158), tourmaline (page 110), and other green gemstones.

---

1 Citrine, heat-treated, rough
2 Citrine, heat-treated, faceted
3 Citrine, heat-treated, rectangle
4 Citrine, natural, rough

5 Citrine, natural, oval
6 Citrine, 2 emerald cuts
7 Prasiolite, rough
8 Prasiolite, 2 faceted stones

Rose Quartz [3–7]

Color: Strong pink, pale pink	Chemical composition: SiO_2 silicon dioxide
Color of streak: White	Transparency: Semi transparent, translucent
Mohs' hardness: 7	Refractive index: 1.544–1.553
Density: 2.65	Double refraction: +0.009
Cleavage: None	Dispersion: None
Fracture: Conchoidal, very brittle	Pleochroism: Absent
Crystal system: (Trigonal) prisms, mostly compact	Absorption: Cannot be evaluated
	Fluorescence: Weak; dark violet

Rose quartz (named after its pink color) is often crackled, usually a little turbid. Coloring agent is titanium. Color can fade. Traces of included rutile needles cause six-rayed stars when cut en cabochon [no. 4]. Deposits are found in Brazil and Madagascar, as well as India, Mozambique, Namibia, Sri Lanka, and the United States. Worked into cabochons, bead necklaces, and ornamental pieces; only the larger clear stones can be faceted [no. 5]. Can be confused with kunzite (page 114), morganite (page 96), and topaz (page 102).

Aventurine [1, 2, 8] Also called Aventurine Quartz

Color: Green, red-brown, gold-brown, aventurescent	Chemical composition: SiO_2 silicon dioxide
Color of streak: White	Transparency: Translucent, opaque
Mohs' hardness: 7	Refractive index: 1.544–1.553
Density: 2.64–2.69	Double refraction: +0.009
Cleavage: None	Dispersion: None
Fracture: Conchoidal, brittle	Pleochroism: Absent
Crystal system: (Trigonal) massive	Absorption spectrum: Green a.: 682, 649
	Fluorescence: Green a.: Reddish

A type of glass discovered by chance (Italian—*a ventura*) around 1700 gave the same name to the similar-looking stone. Mostly dark green with metallic glittery appearance caused by included fuchsite (green mica) or red- to gold-brown caused by hematite leaves. Deposits are found in Brazil, India, Austria (Steiermark), Russia (Urals, Siberia), and Tanzania. Used for ornamental objects and cabochons. Can be confused with iridescent analcite (page 212), aventurine feldspar (page 166), emerald (page 90), jade (page 154), as well as aventurine glass.

Prase [9]

Prase is a leek-green (Greek—*prason*) quartz aggregate, usually classified as a chalcedony, whose color is caused by chlorite inclusions. Deposits occur in Saxony (Germany), Finland, Austria (Salzburg), and Scotland. Can be confused with amazonite (page 164) and jade (page 154).

Blue Quartz [10]

This is a coarse-grained, turbid-blue quartz aggregate (quartzite). The inclusions of crocidolite fibers cause the color. Deposits are found in Brazil, Austria (Salzburg), Scandinavia, South Africa, and Virginia. Used for ornaments. Can be confused with dumortierite quartz (page 182) and lapis lazuli (page 172).

1 Aventurine, 5 cabochons	6 Rose quartz, 6 cabochons
2 Aventurine, rough, partly polished	7 Rose quartz, baroque neclace
3 Rose quartz, rough	8 Aventurine, rough, partly polished
4 Star rose quartz, 20.23ct	9 Prase, 2 cabochons
5 Rose quartz, octagon, 8.16ct	10 Blue quartz, rough, partly polished

Quartz Cat's-Eye [1, 2]

Color: White, gray, green, yellow, brown	Transparency: Semitransparent translucent
Color of streak: White	Refractive index: 1.534–1.540
Mohs' hardness: 7	Double refraction: None
Density: 2.58–2.64	Dispersion: None
Cleavage: None	Pleochroism: Absent
Fracture: Irregular	Absorption: Not diagnostic
Crystal system: (Trigonal) usually massive	Fluorescence: None
Chemical composition: SiO_2 silicon dioxide	

Quartz cat's-eye is quartz in which numerous fiber-like inclusions of rutile create chatoyancy. Sensitive to some acids. Deposits found in Sri Lanka, as well as Brazil and India. Cut en cabochon, shows chatoyancy like a cat's-eye (hence the name).

Can be confused with chrysoberyl cat's-eye (page 98). Decolorized hawk's- and tiger's-eye are sometimes substituted for quartz cat's-eye. Syntheses are known. The name "cat's-eye" (without the word "quartz") is taken to mean only chrysoberyl cat's-eye (page 98).

Hawk's-Eye [3, 4]

Finely fibrous, opaque aggregate formed when quartz replaces the mineral crocidolite (type of asbestos), blue-gray to blue-green. Iridescence of planes; fractures have silky luster. It is sensitive to some acids. Found associated with tiger's-eye (see below). Used for ornamental objects and costume jewelry. Cabochons show chatoyancy (small ray of light on surface) which is reminiscent of the eye of a bird of prey. Even flat pieces show a similar effect.

Tiger's-Eye [5, 6]

Color: Gold-yellow, gold-brown	Transparency: Opaque
Color of streak: Yellow-brown	Refractive index: 1.534–1.540
Mohs' hardness: 6½–7	Double refraction: None
Density: 2.58–2.64	Dispersion: None
Cleavage: None	Pleochroism: Absent
Fracture: Fibrous	Absorption: Not diagnostic
Crystal system: (Trigonal) fibrous aggregate	Fluorescence: None
Chemical composition: SiO_2 silicon dioxide	

Formed from hawk's-eye (see above) where the iron from the decomposed crocidolite has oxidized to a brown color, keeping the fibrous structure. The luster is silky on fractures. Typically displays chatoyant stripes, because structural fibers are crooked or bent. It is sensitive to some acids. Found together with hawk's-eye in slabs of a few inches thickness. Most important deposits are found in South Africa (where the export of raw material is forbidden), also in Australia, Burma (Myanmar), India, Namibia, and the United States (California). Used for necklaces, costume jewelry, and objets d'art. When cut en cabochon, the surface shows chatoyancy reminiscent of the eyes of a cat. Red tiger's-eye is artificially dyed. (For tiger's-iron matrix, see page 202.)

1 Quartz cat's-eye	4 Hawk's-eye, 2 cabochons
2 Quartz, cat's-eye, 3.96ct	5 Tiger's-eye, partly polished
3 Hawk's-eye, rough	6 Tiger's-eye, 7 cabochons

Chalcedony

Chalcedony is used by gemologists as a species name for all cryptocrystalline quartzes (compare page 116, e.g., agate, petrified wood, chrysoprase, bloodstone, jasper, carnelian, moss agate, onyx, and sard) as well as specifically only the bluish-white-gray variety, the actual chalcedony (i.e., chalcedony in the narrow sense). Some scientists ascribe only the fibrous varieties to the chalcedonies; the grainy jasper (page 146) then forms its own group.

Whereas the crystal quartzes (rock crystal, amethyst, etc.) have a vitreous luster, chalcedonies are, in their natural state, waxy or dull. Syntheses which can be used economically are not known.

The following data refers to actual chalcedony (i.e., in the narrow sense).

Color: Bluish, white, gray	Transparency: Dull, translucent
Color of streak: White	Refractive index: 1.530–1.540
Mohs' hardness: 6½–7	Double refraction: up to 0.004
Density: 2.58–2.64	Dispersion: None
Cleavage: None	Pleochroism: Absent
Fracture: Uneven, shell-like	Absorption spectrum: Dyed blue: 690–660, 627
Crystal system: (Trigonal) fibrous aggregates	
Chemical composition: SiO_2 silicon dioxide	Fluorescence: Blue-white

Chalcedony [nos. 4–6], probably named after an ancient town at the Bosporus, consists of microscopic fibers, which are parallel to each other. Chalcedony shows macroscopically radiating, stalactitic, grape-like or kidney shapes [no. 4]. Always porous; can therefore be dyed (compare to page 136). Natural chalcedony normally has no banding. The trade also offers parallel layered, artificial blue-colored agate as chalcedony [no. 5]. Deposits are found in Brazil, India, Madagascar, Namibia, Zimbabwe, Sri Lanka, Uruguay, and California. In ancient times used for cameos; today used in the arts and crafts for rings and necklaces. Can be confused with tanzanite (page 160).

Chrome chalcedony (also called mtorodite or mtorolite)   Trade name for a chalcedony from Zimbabwe, which has a natural green color from chromium.

Carnelian [2, 3]   Chalcedony Species

Carnelian is probably named after the color of the kornel cherry because of its color. It is a brownish red to orange, translucent to opaque chalcedony variety. The coloring agent is iron; the color can be enhanced by heating. Deposits are found in Brazil, India, and Uruguay. Most carnelians offered today are agates which are dyed and then heat-treated. When held against the light, the color variety shows stripes, natural carnelian shows a cloudy distribution of color. Used similarly to other chalcedonies. Can be confused with jasper (page 146).

Sard [1]   Chalcedony Species

Red-brown to brown variety of chalcedony (named after town in Asia Minor). No strict separation from carnelian (darker and browner stones are usually called sard), with common deposits, and uses. Many of the "sards" on the market are dyed.

1 Sard, 8 faceted and cabochon stones
2 Carnelian, rough
3 Carnelian, 7 tablet and cabochon stones

4 Chalcedony nodule, partly polished
5 Chalcedony, 3 banded stones
6 Chalcedony, 7 cabochons

Chrysoprase [1–4]

Color: Green, apple-green	Transparency: Translucent, opaque
Color of streak: White	Refractive index: 1.530–1.540
Mohs' hardness: 6½–7	Double refraction: up to 0.004
Density: 2.58–2.64	Dispersion: None
Cleavage: None	Pleochroism: Absent
Fracture: Rough, brittle	Absorption: Natural ch.: 444; Dyed with
Crystal system: (Trigonal) microcrystalline	nickel: 632, 444
aggregates	Fluorescence: None
Chemical composition: SiO_2 silicon dioxide	

Chrysoprase is considered the most valuable stone in the chalcedony group. The name (Greek—gold-leek) seems misapplied today. The microscopic fine quartz fibers have a radial structure. The coloring agent is nickel. Large broken pieces are often full of fissures with irregular colors. Color can fade in sunlight and when heated (be careful when soldering). Colors may recover under moist storage.

Occurs as nodules or fillings of clefts in serpentine rocks and in weathered materials of nickel ore deposits. Long ago, the deposit of Frankenstein (Zabkowice) in Upper Silesia, Poland, was the most important mine, but it has been worked out since the 14th century. Today's deposits include Australia (New South Wales), Brazil, India, Kazakhstan, Madagascar, Russia (the Urals), Zimbabwe, South Africa, Tanzania, and California.

Used as cabochons, for necklaces, and for ornamental objects. In earlier centuries, it was used as a luxurious decorative stone for interior decoration, such as in the Wenceslaus Chapel in Prague and in Sanssouci Castle in Potsdam (near Berlin).

Possibilities for Confusion   With chrome chalcedony (page 126), jade (page 154), prasopal (page 152), prehnite (page 188), smithsonite (page 198), variscite (page 196), and artificially colored green chalcedony.

Chrysoprase matrix [3, 4]   Chrysoprase with brown or white matrix rock. Used for ornamental objects and in jewelry when cut en cabochon.

Bloodstone [5, 6]

Bloodstone is an opaque, dark-green chalcedony with red spots. An old name still used in Europe is heliotrope (Greek—sun turner). Particles of chlorite or included hornblende needles cause the green color. Red spots are caused by iron oxide. The colors are not always constant. The most important deposits are in India, also in Australia, Brazil, China, and the United States. Used often as seals for men's rings and for other ornamental objects.

In the trade, the term *blood jasper* is sometimes used. Bloodstone, however, is not a jasper at all (page 146), even though a radial structure with spherical aggregates can simulate a grainy appearance. Other data as for chrysoprase.

---

1 Chrysoprase, 2 pieces, partly polished	4 Chrysoprase-matrix, partly polished
2 Chrysoprase, 4 cabochons	5 Heliotrope, rough, partly polished
3 Chrysoprase, 2 stones with matrix	6 Heliotrope, 7 tablet and cabochon
	stones

Dendritic Agate [1-4] Also called Mocha Stone   Chalcedony Species

Dendritic agate is a colorless or whitish-gray, translucent chalcedony with tree- or fern-like markings, the dendrites (Greek—tree-like). According to some authorities, the term agate (compare page 132) is not strictly correct, as there are no bandings but the term is not generally considered deceptive. The dendrites are iron or manganese inclusions of brown or black color. Despite their looks, they have nothing to do with the organic world, but rather they resemble ice crystals on windows in the winter. They are formed at very fine fracture surfaces through crystallization of weathered solutions of neighboring rock.

They are found with other chalcedonies. Important deposits are in Brazil (Rio Grande do Sul); also in India and the United States. Because the Indian stones formerly came via the Arabian harbor of Mocha, these stones are also called mocha stones.

Scenic agate [2]   A dendritic agate where the included dedrites resemble landscape-like images in brown or reddish color tones.

Mosquito stone [4] (also called midge stone)   A dendritic agate where the dendrites do not hang together but show ball-like growths, reminiscent of swarms of mosquitoes.

Used to make rings, brooches, and pendants. Because the colorful inclusions are at various depths of the rough stone, the cutter must try to bring the image-like markings closer to the surface by taking off the upper layers. Inevitably, a not quite even surface-form may result. Imitations have been attempted with silver nitrate.

Moss Agate [5, 6]   Chalcedony Species

Color: Colorless with green, brown, or red inclusions	Transparency: Translucent
Color of streak: White	Refractive index: 1.530-1.540
Mohs' hardness: 6½–7	Double refraction: up to 0.004
Density: 2.58-2.64	Dispersion: None
Cleavage: None	Pleochroism: Absent
Fracture: Rough	Absorption: Cannot be evaluated
Crystal system: (Trigonal) microcrystalline	Fluorescence: Variable, cannot be used for a diagnosis
Chemical composition: SiO_2 silicon dioxide	

Moss agate is a colorless, translucent chalcedony with inclusions of green hornblende (or chlorite) in moss-like patterns (therefore the name). Moss agate colors are brown and red through oxidation of the iron hornblende. The name agate (page 132) is generally accepted even though the stone is not banded.

It occurs as filler in fissures or as pebbles. India supplies the best quality specimens. Other deposits are in China, Russia (the Urals), and Colorado.

It is used in thin slabs, so that the moss-like image can be seen effectively, especially as plates, cabochons for rings, brooches, and pendants, as well as for ornamental objects.

Possibilities for Confusion   With other natural gemstones hardly exist due to their characteristic looks. But there are good imitations by doublets.

---

1 Dendritic agate, fern-like
2 Scenic agate
3 Dendritic agate, 2 pieces, radial inclusions

4 Mosquito agate
5 Moss agate, 10 cabochons
6 Moss agate, 2 pieces, partly polished

Color: All colors, banded	Transparency: Translucent, opaque
Color of streak: White	Refractive index: 1.530–1.540
Mohs' hardness: 6½–7	Double refraction: up to 0.004
Density: 2.60–2.64	Dispersion: None
Cleavage: None	Pleochroism: Absent
Fracture: Uneven	Absorption spectrum: Dyed green: 700,
Crystal system: (Trigonal) microcrystalline	(665), (634)
aggregates	Fluorescence: Varies with bands: partly
Chemical composition: SiO₂ silicon dioxide	strong; yellow, blue-white

Agate is a banded, concentric shell-like chalcedony, sometimes containing opal substance (see page 150). The fine quartz fibers are oriented vertically to the surface of the individual band layers. The bands can be multicolored or of the same color. The agates of the exhausted German mines had soft to strong colors (especially pink, red, or brownish) and were separated by bright gray bands. The South American agates are mostly dull gray and without special markings; only through dyeing (page 136) do they receive their lively colors. Transparency of agates varies from nearly transparent to opaque. In thin slabs, even the opaque agates are mostly translucent. The name agate is supposedly derived from the river Achates (now called the Drillo) in Sicily.

Origin Agates are found as ball- or almond-shaped nodules with sizes ranging from a fraction of an inch to a circumference of several yards; more rarely they are found as fillings of crevices in volcanic rocks (such as melaphyre and porphyry). The bands are thought to be formed by rhythmic crystallization, but scientific opinions vary as to how. It was thought that the agate bands crystallize gradually in hollows formed by gas bubbles from a siliceous solution. Recently, the theory that their formation is simultaneous with that of the matrix rock has won support. According to this idea, the liquid drops of the silicic acid cool with the cooling rock and produce a layered crystallization from the outside. A new theory postulates that rather than the liquids penetrating the agate walls, colloid solutions, i.e., substances with very fine sizes of grains, flow into the agate hollows. The various agate bands vary in thickness, but normally their thicknesses remain constant throughout the nodule. Where the inner cavity of the nodule is not filled with an agate mass, well-developed crystals may have formed: rock crystal (page 116), amethyst (page 118), and smoky quartz (page 116); sometimes accompanied by *Anhydritspat* (page 206), *Ankerit* (page 68), Baryt (page 206), calcite (page 208), Goethit, hematite (page 162), siderite (page 206), and zeolite. A nodule with crystals in the central cavity if called a druse [no. 5]; if the inside is completely filled, one speaks of a geode [2].

For further information, see the following:

---

1 Agate nodule in cross section, banded agate, ⅓ natural size, Uruguay
2 Agate nodule in cross section, orbicular agate with concentric eye (eye agate), ⅔ natural size, India
3 Agate nodule in cross section, orbicular, ⅔ natural size, India
4 Agate nodule in cross section, orbicular agate with eccentric eye (eye agate), ⅔ natural size, India
5 Agate geode, ⅔ natural size, Brazil

Agate Varieties According to sample, design, or structure of the agate layer, there are numerous trade names.

Eye agate [page 133, nos. 2, 4] Ring-shaped design with point in the center, similar to an eye. A type of orbicular agate.

Layer agate [page 133, no. 1] Layers/bands of about the same size parallel to the outer wall of the agate nodule.

Dendritic agate [page 131, no. 1-4] Colorless or whitish, translucent chalcedony with dendrites.

Enhydritic agate (also called enhydro or water stone) Agate nodule or mono-colored chalcedony nodule, partly filled with water which can be seen through the walls. After the agate is taken from surrounding rock, the water often dries out.

Fortification agate [page 135, no. 2] Agate bands with jutting-out corners like the bastions of old fortresses.

Fire agate Opaque, limonite-bearing layered chalcedony with iridescence which is created through diffraction of the light by the layered structure.

Orbicular agate [page 133, no 3] Circles of the agate layers, arranged concentrically or excentrically around a centerpoint.

Moss agate [page 131, nos. 5, 6] Translucent chalcedony with moss-like inclusions of hornblende or chlorite.

Scenic agate [page 131, no. 2] Shows landscape-like images through dendrites.

Pseudo-agate (also called polyhedric quartz) [page 137, no. 2] Interior similar to agate with layering and druse opening, although outside not nodule-like, but geometric shape. Formed probably as wedge-filling of crystals, which later were loosened. Occurrences in loose rocks in Brazil. Individual pieces can measure up to 28 in (75 cm). Many fanciful names.

Tubular agate [page 135, no. 3] Agate with numerous tubes (old feeding canals). The mouth-opening of the canals is usually bordered concentrically.

Thunder egg or Sandstone [page 135, no. 1] Layered agate nodule with strongly furrowed outer surface. Deposits in Oregon (United States) and Mexico.

Brecciated agate [page 137, no. 3] Agate, broken, but cemented together by quartz.

Deposits The most important agate deposits at the beginning of the 19th century in the neighborhood of Idar-Oberstein/Rhineland-Palatinate, Germany. These have now been worked out. Nodules were found as large as a human head with beautiful colors of gray, pink, red, yellow, brown, and pale blue. These could not be dyed.

---

1 Cross section of agate nodule, sard stone with concentric, shell-like outer layers and parallel layers on the inside; ⅓ natural size. Found in Brazil

2 Cross section of agate nodule, fortification agate with agate layers having corners that jut out like bastions; ⅓ natural size. Found in Brazil

3 Agate nodule partially cut, tubular agate with numerous canals, former feeding tubes; ½ natural size. Found in Idar-Oberstein/Rhineland-Palatinate, Germany

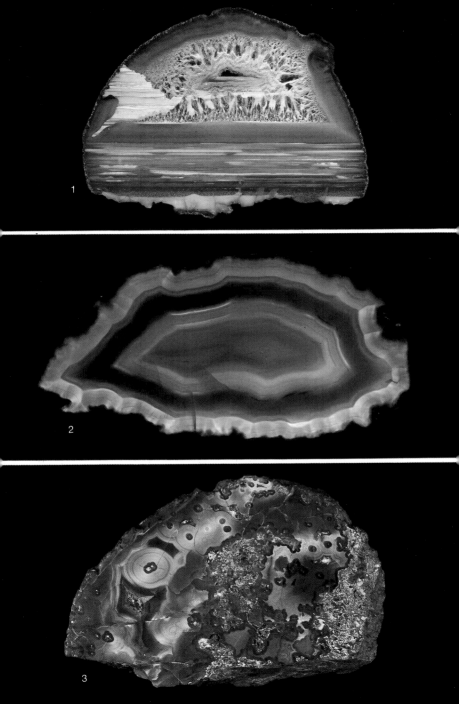

The most important deposits today are in the south of Brazil (Rio Grande do Sul) and in the north of Uruguay. The deposits are layered in weathered materials and river sediments, and are derived from melaphyric rocks. The color is generally gray; the striations are hardly recognizable. They can be given an attractive appearance by dyeing (see below). Other deposits are in Australia (Queensland), China, India, the Caucasus, Madagascar, Mexico, Mongolia, Namibia, Wyoming, and Montana.

Agate Coloring The South American deposits, the most important suppliers of agate overall, produce agates which normally appear opaque gray and without markings. Only when dyed do they obtain their coloring and lively patterns.

The art of dyeing was known to the Romans. In Idar-Oberstein/Rhineland-Palatinate, it has been practiced since the 1820s and brought to a perfection not achieved anywhere else. This is the reason why the town has developed into the most important center for the cutting of agate and other stones.

The absorption of the dye varies with the porosity and water content of the individual layer of the agate. White layers, consisting of dense quartz aggregates, absorb little or no color. Layers which are easily dyed are called soft, the others hard.

Details of the process are a commercial secret. Generally, inorganic pigments are used, as the organic ones tend to fade in light and are less intense. Dyed agates are not recognizable with the naked eye, unless a certain color, such as deep blue, does not exist in natural agate. Although agate dyeing is considered routine in the trade, US FTC guidelines and many industry organizations say it should be disclosed (see page 29).

Before they are dyed, the agates are cleaned with a warm acid or a caustic solution, cut into their final form, and sometimes even polished.

Coloring red [1b]   Pigment is iron oxide. The agate is laid in a solution of iron nitrate, then strongly heated. By altering techniques, various reds can be obtained. Natural yellow layers turn red by heating alone.

Coloring yellow [1b]   Pigment is iron chloride. The agate is first saturated with hydrochloric acid, then slightly warmed, producing a lemon-yellow color.

Coloring brown [1b]   By treatment with sugar solution and heating, browns are produced. By use of cobalt nitrate, supposedly the same results can be achieved.

Coloring black [1c]   Pigment is carbon. Use of concentrated honey or sugar solution, following treatment with heated sulfuric acid, produces a dark black color in agate. By certain variations, browns can be produced. Cobalt nitrate is said to have the same effect.

Coloring green [1d]   Pigment is iron. Saturation of the agate with chromium salt solution and subsequent heat treatment produces green. The use of nickel nitrate solution and heat is said to produce the same result.

Coloring blue [1e]   Pigment is iron. Agates are first placed in a solution of yellow potassium ferro-cyanide and subsequently boiled in hydrous iron sulfate. This produces "Berliner" blue.

1 Agate plate, a—natural, b–e—colored; ⅓ natural size, Brazil
2 Pseudo-agate, ½ natural size, Brazil (compare with page 134)
3 Brecciated agate, partly polished; ½ natural size, U.S. (compare with page 134)

Old agate mill with waterwheel, Idar-Oberstein/Rhineland-Palatinate, Germany.

Development of the agate industry in Idar-Oberstein Agate has a special position among gemstones; there is a unique industry centered around it in Idar-Oberstein in Rhineland-Palatinate, Germany. The bases for this development were favorable natural occurrences, for instance some finds of agate and jasper, good local sandstone for the production of cutting and polishing wheels, and water power to work the wheels.

Gemstones have been worked in and near Idar-Oberstein since the first part of the 16th century. Agate polishing at the Idar-brook was first mentioned in documents in 1548, but it is known that, one hundred years before that, agate, jasper, and quartz were locally mined, but possibly worked somewhere else.

Toward the end of the 17th century, there were about 15 workshops; and around 1800, 30 workshops were cutting agate and using the river for energy. Toward the beginning of the 19th century, local agate deposits were beginning to be worked out, and many experts left the area. However, new life was brought to the industry when, by chance, some emigrants who were wandering around as musicians discovered large agate deposits in Brazil. By 1834 the first supply of Brazilian agate had reached Idar-Oberstein, and by 1867 there were 153 polishing shops.

With the emergence of steam power and especially since the advent of electric energy, the industry has been decentralized. Today there are numerous workshops in and around the district.

Now that more and more countries around the world have established their own gemstone industries, Idar-Oberstein has found its place in specializing on processing high-quality agates almost exclusively.

---

1 Bowl of Brazilian agate, diameter 4.2 in/11.2 cm, height 2.6 in/6.5 cm
2 Bowl of Brazilian agate, diameter 5¼ in/14 cm, height 1½ in/4 cm

Agate polishers at work toward the end of the 19th century; Idar-Oberstein/Rhineland-Palatinate, Germany.

The history of agate polishing The oldest way to polish agate was to rub it against a horizontal sandstone. Polishing against a vertically rotating sandstone probably became common during the 14th century. A waterwheel outside the house was turned by a river or dammed pond to drive the axle inside the house. Several sandstone grinding wheels were mounted vertically onto this axle. These wheels were about 56 in (150 cm) high and 15–19 in (40–50 cm) wide. The polishers lay on their stomachs on special chairs and pushed the agate hard against the rotating wheels, which were sprinkled with water constantly. Because the sandstone wheel had an obtuse angle in the middle of its working surface, two workers could share one wheel.

With the increasing use of steam and, later, electricity and also the new method of polishing using carborundum, a sitting posture was adopted. It is more comfortable and does not require as much effort as formerly.

For further information, see modern polishing on page 60.

Uses At least 3,000 years ago, agate was used as a gemstone by the Egyptians. Today it is used for objets d'art, decorative purposes, beads, rings, brooches, pendants, and as layer stones for cameos (page 142). It is also used by industry because of its toughness and resistance to chemicals.

1 Agate, decorative egg	7 Agate, handle for a letter opener
2 Agate, pendant	8 Agate, brooch
3 Agate, ring	9 Agate, seal
4 Agate, mortar	10 Agate, pill box
5 Agate, handles for knives and forks	11 Agate, dentist's instrument
6 Agate, handles for a manicure set	12 Agate, letter opener

Layer Stones   Layer stones are multilayered materials used in the art of gem carving and engraving, also called glyptography. Usually this material is cut from agates or other chalcedonies with even parallel layers, a lighter layer above a darker one. Brazil supplies the best raw material, usually two-layered, but sometimes three-layered ones are seen. Some masterpieces are cut out of five-layered material. Engravings in multilayered and curved agates are rare.

Some explanations concerning layer stones and engraving work follow.

Onyx   Layer stone with the combination of a black base and a white upper layer, also called true onyx or Arabic onyx [page 145, no. 4]. In cabochons and beads, black and white layers may alternate. On the other hand, onyx is also a name sometimes used for unicolored chalcedony (e.g., black onyx). This must not be confused with onyx marble, which in short is also called onyx (compare with page 218). The name onyx has its origin in the Greek language and means "fingernail," probably because of its weak transparency.

Sard onyx   Layer stone with brown base, upper layer white [page 143, no. 1].

Cornelian onyx   Layer stone with red base, upper layer white [page 145, no. 5].

Niccolo   Layer stone whose upper layer is very thin, which results in translucent blue-gray color tones due to diffusion of the light and showing through of the black base. Popular as seal rings for engraving coats-of-arms and monograms.

Intaglio (Italian)   Name for engraving with negative image, as used for seals.

Cameo (Italian)   Name for a relief which is cut so that it is raised.

The layers in agate or chalcedony required for this work are not often found in nature in the colors of onyx, carnelian, or sard. Therefore such stones are usually dyed, as described in more detail on page 136. The natural and the dyed stones have the same names.

Recently onyx layer stones have been produced from unlayered, unicolored gray chalcedony, which can be dyed, as follows: A square block is saturated with a solution of cobalt chlorate and chlorammonium, and is thus dyed black. With the help of hydrochloric acid, this color is then removed up to a depth of 0.04 in (1 mm). When the block is sawn in half, the sawn surfaces are black and the reverse sides are white; however, it is said that the dark color tends to fade.

Creation of a Cameo

1 A piece of layer stone with different colors in even parallel layers is cut out or broken out of a block of agate, as seen in the background.

2 Several two-layered stones can be cut out of the first piece.

3 The lower layer is dyed black or red-brown. The upper layer remains white because it does not absorb the pigment.

4 The main features of the cameo are indicated. In mass production, a template is used.

5–8 These illustrate the art of the engraver; his experience, knowledge of the stone, and his technical perfection lend the personal touch.

9 The final result is an example of the masterly precision of the engraver's art.

142

A gem engraver at work

The technique of stone engraving The main tool of the engraver is a small lathe with a horizontal spindle, to which, according to need, various instruments can be attached. These can be wheels, spheres, cones, or needles, which are kept handy on a rack nearby. The spindle is driven by an electric motor at 3000–5000 r.p.m. The spindle is rigid and the engraver guides the stone by hand. This requires great precision and knowledge of the particular stone. A working period lasts no more than two to three hours.

The rotating tips are prepared with diamond polishing powder and oil; by this means, they are cooled and given an abrasive surface because the tiny diamond particles are pressed into the softer iron of the instruments during the engraving process.

The polishing is performed with wood, leather, or another softer material using water and special polishing pastes. This process also removes any marks made by a metal pencil during preliminary sketching.

Recently, flexible spindles are also available. However, they are usually used only for larger sculptures, where the stone is too heavy to be guided by hand.

Even though nowadays all gemstones are used for engravings, agate, especially layer stones, are worked most of all.

1 Abstract modern engraving, onyx	5 Girl with blossom, carnelian onyx
2 Abstract modern engraving, onyx	6 Coat-of-arms, onyx, Brazil
3 Shadow cameo, onyx, Brazil	7 Shaded engraving, onyx, Brazil
4 Woman's head, Parisian style, onyx	8 Initials, onyx, Brazil

Color: All colors, mostly striped or spotted	Transparency: Opaque, even in thin slabs opaque
Color of streak: White, yellow, brown, red	Refractive index: About 1.54
Mohs' hardness: 6½–7	Double refraction: None
Density: 2.58–2.91	Dispersion: None
Cleavage: None	Pleochroism: Absent
Fracture: Splintery, conchoidal	Absorption: Cannot be evaluated
Crystal system: (Trigonal) microcrystalline aggregate	Fluorescence: None
Chemical composition: SiO_2 silicon dioxide	

Jasper is usually considered as chalcedony (page 126); sometimes, however, scientists put it in a group by itself within the quartz group because of its grainy structure.

The name jasper is derived from the Greek and means "spotted stone." In antiquity, however, jasper referred to completely different stones than today, i.e., to green, transparent varieties.

The finely grained, dense jasper contains up to 20 percent foreign materials, which determine its color, streak, and appearance. Uniformly colored jasper is rare; usually it is multicolored, striped, or flamed. Sometimes jasper can be grown together with agate or opal. There is also fossilized material (page 148).

Occurs as fillings of crevices or fissures or in nodules. Deposits are found in Egypt, Australia, Brazil, India, Canada, Kazakhstan, Madagascar, Russia, Uruguay, and the United States. Used for ornamental objects, cabochons, and for stone mosaics. Care must be taken during cutting and polishing; banded jasper tends to separate along the layers.

Varieties According to color, appearance, occurrence, or composition, there are many names used in the trade.

Agate jasper (jaspagate) Yellow, brown, or green blended, grown together with agate.

Egyptian jasper (Nile pebble) Strongly yellow and red.

Banded jasper [13] Layered structure with more or less wide bands.

Basanite (touchstone) Fine-grained, black. Is used by jewelers and goldsmiths for streak-tests of precious metals.

Blood jasper Name sometimes used for bloodstone (page 128).

Hornstone (chert) Very fine grained, gray, brown-red, more rarely green or black. Sometimes hornstone is understood as a general synonym for jasper.

Scenic jasper Brown marking, caused by iron oxide, resembling a landscape.

Moukaite [4] Pink to light red, cloudy. Found in Australia.

Nunkirchner jasper Whitish gray, rarely yellow or brownish-red (named after deposit in Rhineland-Palatinate). Dyed with Berliner blue, misleadingly called "German lapis" or "Swiss lapis" in imitation of lapis lazuli (page 172).

Plasma Dark green, sometimes with white or yellow spots.

Silex [14] Yellow and brown-red spotted or striped.

1 Jasper breccia, Australia	8 Yellow jasper, cabochon, Australia
2 Pop jasper, 2 stones, South Africa	9 Yellow jasper, Australia
3 Pop jasper, 2 stones, Australia	10 Multicolored jasper, India
4 Moukaite, Australia	11 Striped jasper, South Africa
5 Multicolored jasper, 2 stones, India	12 Multicolored jasper, India
6 Multicolored jasper, cabochon, Australia	13 Banded jasper, Australia
7 Zebra jasper, South Africa	14 Silex, Egypt
	15 Multicolored jasper, India
	16 Multicolored jasper, rough, India

Petrified Wood Also called Fossilized Wood   Chalcedony Species

Color: Brown, gray, red	Chemical composition: SiO_2 silicon dioxide
Color of streak: White, partly colored	Transparency: Opaque, even in thin slabs
Mohs' hardness: 6½–7	Refractive index: About 1.54
Density: 2.58–2.91	Double refraction: Weak or none
Cleavage: None	Dispersion: None
Fracture: Uneven, splintery	Pleochroism: Absent
Crystal system: (Trigonal) microcrystalline aggregate	Absorption spectrum: Cannot be evaluated
	Fluorescence: None

Petrified wood is fossilized wood with the mineral composition of jasper, chalcedony, and, less frequently, opal; it consists of silicon dioxide only. The wood has not actually become stone as is usually understood by the layman. The organic wood is not changed into stone, but only the shape and structural elements of the wood are preserved. The expert speaks of a pseudomorphosis of chalcedony (or jasper or opal) after wood.

Well-preserved petrification occurs only where trees after their death are quickly covered with fine-grained sedimentary rock. Thus the outer structure of the wood is preserved in a negative form within the enclosing rock. Circulating waters loosen and decompose the organic substances and replace them with mineral substances. Therefore, it is not a *change* that takes place, but rather an *exchange*. Sometimes this process is successive (from removal and adding of substances) so that the inner structural elements of the wood, the annuals rings [no. 5], the structure of the cells, even wormholes, are preserved. It can also happen that the appearance is totally changed by the crystallization process.

The colors are mostly dull gray or brown, sometimes also red, pink, light brown, yellow, and even blue to violet. The colors become stronger with cutting and polishing.

The most important occurrence is the "petrified wood" near Holbrook in Arizona (United States). There are fossilized tree trunks of up to 213 ft (65 m) long and 10 ft (3 m) thick belonging to the araucaria variety of plants. The tree trunks were deposited there from various parts by water about 200 million years ago, and then covered by several hundred yards of sediment. In the course of time, part of the fossilized wood was exposed by weathering from the enclosing sandstone. Nowhere is the fossilized wood as splendidly colored as in Arizona. In order to preserve this unique natural beauty spot, the "Petrified Forest" was declared a national park in 1962. That also means that no visitor is allowed to take a piece of these petrified materials as souvenirs.

There are smaller deposits of petrified wood on all the continents. Egypt supplies good quality (Dschel Moka Ham near Cairo), as does Argentina (Patagonia), Canada (Alberta), and Wyoming. In Nevada (Virgin Valley), the fossilized wood shows the beautiful color play of opal.

It is mostly used for ornamental objects and decorative pieces (tabletops, ashtrays, bookends, paperweights), less frequently for jewelry purposes.

---

1 Petrified wood, ashtray
2 Petrified wood, bottle cork holder
3 Petrified wood, five fern-tree
   pieces

4 Petrified wood, partly polished
5 Petrified wood, with year rings
6 Petrified wood, two sections of
   tree trunk

The illustrations are 50 percent smaller than the originals.

148

Opal Species

The name is derived from an Indian (Sanskrit) word for "stone." It is divided into three subgroups: the precious opals, the yellow-red fire opals, and the common opals. Their physical properties vary considerably.

Color: All colors, partially play-of-color
Color of streak: White
Mohs' hardness: 5½–6½
Density: 1.98–2.50
Cleavage: None
Fracture: Conchoidal, splintery, brittle
Crystal system: Amorphous; kidney- or grape-shaped aggregates
Chemical composition: $SiO_2 \bullet nH_2O$ hydrous silicon dioxide

Transparency: Transparent, opaque
Refractive index: 1.37–1.52
Double refraction: None
Dispersion: None
Pleochroism Absent
Absorption spectrum: Fire opal: 700–640, 590–400
Fluorescence: White o.: white, bluish, brownish, greenish
Fire o.: greenish to brown

Precious Opal

The special characteristic of these gems is their play-of-color, a display of rainbow-like hues which (especially in rounded cut forms) changes with the angle of observation. The electron-microscope, using a magnification of 20,000, reveals the cause: tiny spheres (as small as 0.001 millimeter in diameter) of the mineral cristobalite layered in siliceous jelly cause the diffraction and interference patterns.

Opal always contains water (3 to 30 percent). It can happen that in the course of time, the stone loses water, cracks, and the play-of-color diminishes. This can, at least temporarily, be restored by saturation with oil, epoxy resin, or water. The aging process is avoided when stored in moist absorbent cotton wool. Care must be taken during setting. A little heat can evaporate the water. Opal is also sensitive to pressure and knocks as well as being affected by acids and alkalies. Opal is often impregnated with plastic to improve appearance.

White opal [11-16]   A precious opal of white or otherwise light basic color with color play.

Black opal [4, 5 and 7–10]   Precious opal with dark gray, dark blue, dark green, and gray-black basic color and play-of-color. Deep black is an exception. Black opals are rarer than white opals.

Opal matrix [1, 2, 6]   Banded growth or leafed inclusion of precious opal with and/or in the matrix rock.

Boulder opal   Precious opal with dark base surface, color play, and high density. Occurs as pebble rock, where opal fills hollows (see picture on page 29).

Harlequin opal   Transparent to translucent precious opal with effective mosaic-like color patterns. Counted among the most desirable opals.

Jelly opal   Bluish-gray precious opal with little play-of-color.

Crystal opal   Transparent with strong color play on colorless, vitreous surface.

1 White opal in matrix	9 Black opal doublet, 16.90ct
2 Black opal in matrix	10 Black opal, 2 doublets
3 Opalized snail	11 White opal, 4 cabochons
4 Black opal, diverse shapes	12 White opal, rough, partly polished
5 Black opal, 86ct	13 White opal, cabochon, 10.39ct
6 Opal matrix, pendant	14 White opal, cabochon, 33.75ct
7 Black opal, 2 triplets	15 White opal, 2 cabochons, 7.78ct
8 Black opal, 4 cabochons	16 White opal, 4 cabochons, 14.21ct

About 20 percent smaller than original. Stones not numbered belong to no. 4.

Deposits   Up to the end of the 19th century, the andesite lavas in the east of Slovakia supplied the best qualities. Then the Australian deposits were discovered. Famous deposits in New South Wales are at Lightning Ridge and White Cliffs; in South Australia at Coober Pedy and Andamooka. Numerous deposits are also found in Queensland. Most of the 0.04–0.08 in (1–2 mm)-thin opal layers are bedded in sandstone. Further deposits are found in Brazil, Guatemala, Honduras, Indonesia, Japan, Mexico, Russia, Nevada, and Idaho.

Possibilities for Confusion   With ammonite (page 240), labradorite (page 166), mother-of-pearl (page 239), and moonstone (page 164). Very thin pieces of opal are sometimes mounted on a piece of common opal or black chalcedony to create an opal doublet or layer opal. Triplets are also made with a protective top layer of rock crystal. Several good imitations made from glass or plastic are also known. In 1970, a synthesis of white and black opal succeeded.

Fakes are prepared by coloring black or matrix opal in order to liven up the play of color.

Fire Opal [1-7]
Fire opal (named after its orange color) often shows no play-of-color. It is usually milky and turbid. The best qualities are clear and transparent [nos. 3, 4, 6], which makes them suitable for being faceted. They are very sensitive to every stress. Deposits are found in Mexico, as well as Brazil, Guatemala, the United States, and Western Australia.

Possibilities for Confusion   With garnet (page 104), rhodochrosite (page 168).

Common Opal [8-14] (Also called Potch)
Common opal is opaque, rarely translucent, and shows no play-of-color. A wide variety of trade names are used.

Agate opal (opal agate)   Agate with light and dark opal layers.

Angel skin opal   Misleading name for palygorskite, an opaque, whitish- to pink-colored silicate mineral.

Wood opal (zeasite)   Yellowish or brownish opal as petrified wood (page 148)

Honey opal [8]   Honey-yellow, translucent opal.

Hyalite (glass opal, waterstone)   Colorless, water-clear opal with strong sheen.

Hydrophane   A milk opal, which has turned turbid due to the loss of water. Through absorption of water, it can become translucent again and have color play.

Porcelain opal   White, opaque milk opal.

Moss opal [11]   Milk opal with dendrites.

Girasol   Almost colorless, transparent opal with a bluish opalescence (see page 46).

Prase opal (chrysopal) [10]   Apple-green opal. Substitute for chrysoprase (page 128).

Wax opal [12]   Yellow-brown opal with wax-like luster.

---

1 Fire opal, rough, Mexico
2 Fire opal, 5 cabochons, 11.80ct
3 Fire opal, 9 faceted stones, 11.95ct
4 Fire opal, 4 faceted stones, 13.61ct
5 Fire opal, rough, Mexico
6 Fire opal, 3 faceted stones, 5.89ct
7 Fire opal, cabochon and oval,
  24.53ct

8 Honey opal, Western Australia
9 Common opal, 3 stones, Mexico
10 Prase opal, Nevada
11 Moss opal, India
12 Wax opal, rough, Hungary
13 Dendrite opal, rough
14 Liver opal or menilite, rough,
  Hungary

The illustrations are 20 percent smaller than the originals.

Jade (Jadeite and Nephrite)

The name jade goes back to the time of the Spanish conquest of Central and South America and derives from *piedra de ijada*, i.e., hip stone, as it was seen as a protection against and cure for kidney diseases. The corresponding Chinese word *yu* has not been generally accepted.

In 1863 in France, the gemstone, which had been known for 7,000 years, was proved to consist of two separate, distinct minerals, namely jadeite and nephrite. Differentiation between jadeite and nephrite is based on properties, but the term jade is used as a description of both.

In prehistoric times, jade was used in many parts of the world for arms and tools because of its exceptional toughness. Therefore nephrite is sometimes called "axe stone." For over 2,000 years, jade was part of the religious cult in China and mystic figures and other symbols were carved from it. In pre-Columbian Central America, jade was more highly valued than gold. With the Spanish conquest, the high art of jade carving in America came to a sudden end. In China, however, this art was never interrupted. In former times, only nephrite was worked in China, but since about 1750 jadeite imported from Burma (Myanmar) has also been used.

Jadeite Species [9–13]   Jade

Color: Green, also all other colors
Color of streak: White
Mohs' hardness: 6½–7
Density: 3.30–3.38
Fracture: Splintery, brittle
Crystal system: Monoclinic; intergrown, grainy aggregate
Chemical composition: $NaAlSi_2O_6$ sodium aluminum silicate

Transparency: Opaque, translucent
Refractive index: 1.652–1.688
Double refraction: 0.020
Dispersion: None
Pleochroism: Absent
Absorption spectrum: Green j.: 691, 655, 630, (495), 450, 437, 433
Fluorescence: Greenish j.: very weak: whitish glimmer

Jadeite (name derived from jade) is very tough and resistant because of its tight growth of tiny interlocking grains. It occurs in all colors. Fractures are dull and when polished greasy.

Imperial jade A jadeite from Burma (Myanmar), colored emerald-green with chromium translucent to almost transparent. Most desired jade variety of all.

Chloromelanite (jade albite, maw-sit-sit) [nos. 15, 16] A rock composed of kosmochlor (a mineral related to jadeite) combined with varying amounts of jadeite, albite feldspar, and other minerals. Its color is deep green with dark green-black spots or veins caused by chlorite. The source is upper Burma (Myanmar).

---

1 Nephrite, rough, partly polished
2 Jadeite, 2 stones, table cut
3 Jadeite, 6 cabochons
4 Nephrite, 2 navettes, together 7.68ct
5 Nephrite, 3 cabochons
6 Jadeite, cabochon
7 Nephrite, cabochon, Wyoming (U.S.)
8 Nephrite, octagon, cabochon

9 Jadeite, rough
10 Jadeite, flat table
11 Jadeite, 4 qualities
12 Jadeite, 2 drops
13 Jadeite, 3 different cuts
14 Nephrite, cat's eye
15 Chloromelanite, antique cut, 14.32ct
16 Chloromelanite, 4 different cuts

All illustrations are 20 percent smaller than the original.

Yünan-jade   Chinese term for jadeite, named for the province through which it was imported from Burma (Myanmar).

Historically important jadeite deposits are in upper Burma (Myanmar), near Tawmaw, interlayered in serpentine, or in secondary deposits in conglomerates or in river gravel. Other deposits are in China, Japan, Canada, Guatemala, Kazakhstan, Russia (Siberia), and California. (For more about usage, possibilities for confusion, and imitations, see below.)

Much of the jadeite sold today is treated by dyeing or impregnation with wax or plastic-type resins to improve color and appearance.

Nephrite [page 155, nos. 1–8, 14]   Jade

Color: Green, also other colors	$(Si_4O_{11})_2$ basic calcium mag. iron silicate
Color of streak: White	Transparency: Opaque
Mohs' hardness: 6–6½	Refractive index: 1.600–1.627
Density: 2.90–3.03	Double refraction: –0.027, often none
Cleavage: None	Dispersion: None
Fracture: Splintery, sharp edged, brittle	Pleochroism: Absent
Crystal system: Monoclinic; intergrown fine fibrous aggregate	Absorption spectrum: (689), <u>509</u>, 490, 460
	Fluorescence: None
Chemical composition: $Ca_2(Mg,Fe)_5(OH)_2$	

Nephrite (Greek—kidney) is a dense, felt-like fibrous aggregate variety of the actinolite-tremolite mineral series that is even tougher than jadeite. Occurs in all colors, often with a yellow tint. The most valuable is green. The luster is vitreous.

Nephrite is more common than jadeite. No deposits are in New Zealand (greenstone), occuring in serpentine rocks, and as river or beach pebbles. Other deposits are found in Australia, Brazil, China (Sinkiang), Canada, Zimbabwe, Russia, Taiwan, Alaska, and formerly also in Poland.

Uses   Jadeite and nephrite are used as cabochons, in other jewelry, and for vases, both decorative and religious. The main cutting centers are in China as well as Taiwan and Hong Kong.

Possibilities for Confusion   (For jadeite and nephrite) with agalmatolite (page 222), amazonite (page 164), aventurine (page 122), californite (page 186), chrysoprase (page 128), hydrogrossular garnet (page 106), pectolite (page 212), plasma (page 146), prase (page 122), prehnite (page 188), serpentine (page 202), emerald (page 90), smaragdite (page 204), smithsonite (page 198), and verdite (page 240).

There are many imitations made from glass and plastic. Glued triplets and colorations in order to improve the color value are known. A large number of greenish stones are falsely offered as jade in the trade.

Indian jade   Misleading trade name for aventurine (page 122) and for aventurine glass.

Russian jade   Trade name for spinach-green nephrite from the region of Lake Baikal, Russia.

Wyoming jade   Trade name for nephrite from Wyoming, also for a green-colored growth of tremoite with albite.

1 Nephrite, elephant, China	6 Jade, horse, China
2 Jade, necklace, China	7 Jade 3 symbolic figures
3 Jade, cigarette holder	8 Jade, necklace, multicolored
4 Nephrite, necklace, Burma	9 Nephrite, pendant
5 Jade, Buddha, China	10 Chloromelanite, pendant

The illustrations are 20 percent smaller than the originals.

156

Peridot Also called Chrysolite, Olivine

Color: Yellow-green, olive-green, brownish	magnesium iron silicate
Color of streak: White	Transparency: Transparent
Mohs' hardness: 6½–7	Refractive index: 1.650–1.703
Density: 3.28–3.48	Double refraction: +0.036 to +0.038
Cleavage: Indistinct	Dispersion: 0.020 (0.012–0.013)
Fracture: Brittle, small conchoidal	Pleochroism: Very weak; colorless to pale green, lively green, olive-green
Crystal system: Orthorhombic; short, compact prisms, vertically striated	Absorption: 497, 495, 493, 473, 453
Chemical composition: $(Mg,Fe)_2SiO_4$	Fluorescence: None

The name probably derives from the Arabic word *faridat* (gem). The name *chrysolite* (Greek—gold stone) was formerly applied not only to peridot but also to many similarly colored stones. The name commonly used in mineralogy is olivine (because of its olive-green color).

Olivine is a mineral that occurs in the series with the end members forsterite (Mg_2SiO_4) and fayalite (Fe_2SiO_4). It has a vitreous and greasy luster, and is not resistant to acids. It tends to burst under great stress; therefore it is sometimes metal-foiled. Rarities are peridot cat's eye and star peridot.

Historically important deposit was on the Red Sea volcanic island Zabargad (St. John), 188 miles (300 km) east of Aswan, Egypt; it was mined for over 3500 years but forgotten for many centuries, and rediscovered only around 1900. Beautiful material is also found in the serpentine quarries in upper Burma (Myanmar). Other deposits have been found in Australia (Queensland), Brazil (Minas Gerais), China, Kenya, Mexico, Pakistan, Sri Lanka, South Africa, Tanzania, and Arizona. In Europe, deposits are found in Norway, north of Bergen.

Peridot was brought to Central Europe by the crusaders in the Middle Ages, and was often used for ecclesiastical purposes. It was the most popular stone during the Baroque period. Popular are table and step cuts; sometimes also brilliant cut, especially when set in gold.

The largest cut peridot weighs 319ct and was found on the island Zabargad; it is in the Smithsonian Institution in Washington, D.C. In Russia, there are some cut peridots which came out of a meteorite that fell in 1749 in eastern Siberia.

Possibilities for Confusion With chrysoberyl (page 98), demantoid (page 106), diopside (page 190), emerald (page 90), idocrase (page 186), moldavite (page 220), prasiolite (page 120), precious beryl (page 96), prehnite (page 188), sinhalite (page 186), and tourmaline (page 110). A green foil sometimes enhances pale stones. Imitations are often constructed of corundum and spinel syntheses. Evergreen bottle glass can be mistaken for peridot. The strong double refraction of peridot is an important distinguishing mark. In thick stones the doubling of the edges of the lower facets can be clearly seen through the polished table with the naked eye.

1 Peridot, 2 emerald cuts, each 4.65ct
2 Peridot, 2 ovals, 5.67 and 6.38ct
3 Peridot, emerald cuts, 4.14ct
4 Peridot, oval, 12.45ct
5 Peridot, 4 different cuts
6 Peridot, antique cut, 24.02ct
7 Peridot, 5 faceted stones
8 Peridot, 4 faceted stones
9 Peridot, 5 cabochons
10 Peridot, crystals, partly tumbled
The illustrations are 20 percent larger than the originals.

Zoisite Species

The mineral zoisite (named after the collector Zois) was first found in the Sau-Alp mountains in Kärnten, Austria, in 1805. It was originally called saualpite, and gemstone quality specimens have only recently been found. The gemstone members of the group are tanzanite and thulite as well as anyolite. For the time being, only individual specimens and/or small amounts of other transparent zoisite varieties of gemstone quality (colorless, green, brown) are known.

Tanzanite [1-11]   Zoisite Species

Color: Sapphire blue, amethyst, violet	calcium aluminium silicate
Color of streak: White	Transparency: Transparent
Mohs' hardness: 6½–7	Refractive index: 1.691–1.700
Density: 3.35	Double refraction: +0.009
Cleavage: Perfect	Dispersion: 0.030 (0.011)
Fracture: Uneven, brittle	Pleochroism: Very strong; purple, blue, brown, or yellow
Crystal system: Orthorhombic; multifaced prisms, mostly striated	Absorption: 595, 528, 455
Chemical composition: $Ca_2Al_3(SiO_4)_3(OH)$	Fluorescence: None

The name tanzanite (after the East African state of Tanzania) was introduced by the New York jewelers Tiffany & Co. In good quality the color is ultramarine to sapphire blue; in artificial light, it appears more amethyst violet. When heated to 752–932 degrees F (400–500 degrees C), the interfering yellowish and brown tints vanish, and the blue deepens. Tanzanite cat's eyes are also rarely found. It is used in facet cut. The only deposit is in Tanzania near Arusha; it occurs in veins or filling of fissures of gneisses.

Possibilities for Confusion   With amethyst (page 118), iolite (page 180), lazulite (page 192), sapphire (page 86), spinel (page 100), and synthetic corundum. There are glass imitations and doublets made from glass with tanzanite crown or made from two colorless synthetic spinels glued with tanzanite-colored glue.

Thulite [15-16]   Zoisite Species
Dense, opaque pink zoisite variety. Named after the legendary island Thule. Deposits found in Western Australia, Namibia, Norway, and North Carolina. Used as cabochon or as an ornamental stone.

Possibilities for Confusion   With carnelian (page 126), eudialyte (page 212), rhodonite (page 168), and ruby (page 82).

Anyolite [12-14]   Zoisite Species
Green zoisite rock with black hornblende inclusions and large, but mostly opaque, rubies. Named according to the native language of the Massai ("green"). It was first discovered in 1954 in Tanzania. Due to its color contrasts, it is an effective gem and ornamental stone.

Possibilities for Confusion   With chloromelanite (page 154) and tourmaline (page 110).

1 Tanzanite, crystal	9 Tanzanite, cabochon, 8.5ct
2 Tanzanite in host rock	10 Tanzanite, 5 faceted stones
3 Tanzanite, 3 broken crystals	11 Tanzanite, 5 cabochons
4 Tanzanite, pear-shaped, 5.2ct	12 Anyolite with ruby
5 Tanzanite, antique cut, 24.4ct	13 Anyolite, 2 cabochons
6 Tanzanite, oval, 3.5ct	14 Anyolite with ruby
7 Tanzanite, brilliant cut, 6.8ct	15 Thulite, 2 pieces, rough
8 Tanzanite, oval, 3.1ct	16 Thulite, cabochon

Hematite [1-4]

Color: Black, black-gray, brown-red	Transparency: Opaque
Color of streak: Brown-red	Refractive index: 2.940–3.220
Mohs' hardness: 5½–6½	Double refraction: –0.287
Density: 5.12–5.28	Dispersion: None
Cleavage: None	Pleochroism: Absent
Fracture: Conchoidal, uneven, fibrous	Absorption spectrum: Not diagnostic
Crystal system: (Trigonal) mostly platy	Fluorescence: None
Chemical composition: Fe_2O_3 iron oxide	

The name hematite (Greek—blood) (in some countries, also called bloodstone) is derived from the fact that, when cut, the saw coolant becomes colored red. The English name bloodstone, however, is applied to a chalcedony variety (page 128). In mineralogy, well-crystallized hematite varieties are called iron luster, finely crystallized ones red iron ore or red ironstone, and radial aggregates are called red glass head. When cut into very thin plates, hematite is red and transparent; when polished, it it metallic and shiny. Deposits with cuttable material are found in Cumberland, England, as well as in Bangladesh, Brazil, China, New Zealand, Czech Republic (Erzgebirge), Minnesota, and occasionally Elba/Italy. Formerly used as mourning jewelry, but today mainly for rings, bead necklaces, and intaglio (deepened engravings).

Possibilities for Confusion   With cassiterite (page 184), davidite (page 216), magnetite (page 214), neptunite (page 212), pyrolusite (page 216), and wolframite (page 214).

Hematine   Trade name for an imitation of hematite with pressed and sintered iron oxide, from the United States. In contrast to hematite, it is slightly magnetic.

Pyrite [5-10] Also called Fool's Gold

Color: Brass-yellow, gray-yellow	Chemical composition: FeS_2 iron sulphide
Color of streak: Green-black	Transparency: Opaque
Mohs' hardness: 6–6½	Refractive index: Cannot be determined
Density: 5.00–5.20	Double refraction: None
Cleavage: Indistinct	Dispersion: None
Fracture: Conchoidal, uneven, brittle	Pleochroism: Absent
Crystal system: (Cubic) cubes, pentagonal	Absorption spectrum: Not diagnostic
dodecahedra, octahedra	Fluorescence: None

Pyrite (Greek—fire, as it produces sparks when knocked) is wrongly called marcasite in the trade. True marcasite is a mineral, in many ways similar to pyrite, but unsuitable for jewelry, as it sometimes powders in air. Because of its similarity to gold, pyrite is often called fool's gold. Disk-like, radial-fibrous aggregates, so-called pyrite-suns, are known. Metallic shiny. Deposits in Peru, also in Bolivia, Mexico, Romania, Sweden, and Colorado.

Possibilities for Confusion   With chalcopyrite (page 206) and gold (page 208).

---

1 Hematite, radial red glass head	6 Pyrite, cube crystals
2 Hematite, 2 broken crystals	7 Pyrite crystals in matrix rock
3 Hematite, 5 cut stones	8 Pyrite, 4 different crystals
4 Hematite, tablet cut, truncated corners	9 Pyrite aggregate as brooch
5 Pyrite aggregate covered with crystals	10 Pyrite octahedron crystal

162

Feldspar Group

For the feldspars (from German *Feld*, field, and *spalten*, to split), there are two main subgroups that produce gems: the potassium feldspars and the plagioclases (a series from calcium to sodium feldspars). There are numerous gemstone varieties.

Amazonite [1–3] Also called Amazon Stone
Feldspar Group

Color: Green, blue-green	potassium aluminum silicate
Color of streak: White	Transparency: Translucent, opaque
Mohs' hardness: 6–6½	Refractive index: 1.522–1.530
Density: 2.56–2.58	Double refraction: –0.008
Cleavage: Perfect	Dispersion: None
Fracture: Uneven, splintery, brittle	Pleochroism: Absent
Crystal system: Triclinic; prismatic	Absorption spectrum: Not diagnostic
Chemical composition: $KAlSi_3O_8$	Fluorescence: Weak; olive-green

Amazonite (derived from the Amazon) is a mostly opaque, green sodium feldspar. Color distribution is irregular; luster is vitreous; it is sensitive to pressure. Deposits are found in Colorado, Brazil, India, Kenya, Madagascar, Namibia, and Russia.

Possibilities for Confusion   With chrysoprase (page 128), jade (page 154), serpentine (page 202), and turquoise (page 170).

Moonstone [7–10]
Feldspar Group

Color: Colorless, yellow, pale sheen	potassium aluminum silicate
Color of streak: White	Refractive index: 1.518-1.526
Mohs' hardness: 6–6½	Double refraction: –0.008
Density: 2.56–2.59	Dispersion: None
Cleavage: Perfect	Pleochroism: Absent
Fracture: Uneven, conchoidal	Absorption spectrum: Not diagnostic
Crystal system: Monoclinic; prismatic	Fluorescence: Weak; bluish, orange
Chemical composition: $KAlSi_3O_8$	

A potassium feldspar of the orthoclase (adularia) species with white shimmer, similar to moonshine (therefore the name) the so-called adularescence (see page 44). Moonstone cat's-eye is also known. Vitrous luster; sensitive to pressure. Deposits found in Sri Lanka, Burma (Myanmar), Brazil, India, Madagascar, and the United States. Used as cabochons.

Possibilities for Confusion   With chalcedony (page 126), synthetic spinel, and glass imitations. Other moonstones from the feldspar group are also known.

Orthoclase [4–6]
Feldspar Group

Orthoclase (Greek—to break straight) is transparent to opaque, often colorless or champagne-colored potassium feldspar. Vitreous luster. Deposits are found in Madagascar, Burma (Myanmar), and Kenya.

Possibilities for Confusion   With apatite (page 194), chrysoberyl (page 98), citrine (page 120), precious beryl (page 96), prehnite (page 188), topaz (page 102), and zircon (page 108).

1 Amazonite, broken crystal	6 Orthoclase, rough, Kenya
2 Amazonite, rough	7 Moonstone, rough, 2 pieces
3 Amazonite, 6 different cuts	8 Moonstone, 7 cabochons, India
4 Orthoclase, broken crystal	9 Moonstone, 2 cabochons, Sri Lanka
5 Orthoclase, 3 faceted stones	10 Moonstone, 3 cabochons, 13.23ct

The illustrations are 10 percent smaller than the originals.

Labradorite [1-4]   Feldspar Group

Color: Dark gray to gray-black, with colorful iridescence; also colorless, brownish	Chemical composition: NaAlSi$_3$O$_8$ to CaAl$_2$Si$_2$O$_8$ sodium calcium aluminum silicate
Color of streak: White	Transparency: Transparent to opaque
Mohs' hardness: 6–6½	Refractive index: 1.559–1.570
Density: 2.65–2.75	Double refraction: +0.008 to + 0.010
Cleavage: Perfect	Dispersion: 0.019 (0.010)
Fracture: Uneven, splintery, brittle	Absorption spectrum: Not diagnostic
Crystal system: (Triclinic), platy, prismatic	Fluorescence: Yellow striations

Labradorite (named after peninsula of Labrador in Canada, where it was found) is a plagioclase feldspar. It shows a schiller (labradorescence) in lustrous metallic tints, often blue and green, although specimens with the complete spectrum are most appreciated. It is mainly caused by interference of light from lattice distortions resulting from alternating microscopic exsolution lamellae of high- and low-calcium plagioclase phases. Vitreous luster; sensitive to pressure. Deposits are found in Canada (Labrador, Newfoundland), also in Australia (New South Wales), Madagascar, Mexico, Russia, and the United States.

Used for bead necklaces, brooches, rings, and ornamental objects. Colorless and yellowish-brown transparent labradorites [no. 3] are cut with facets.

Spectrolite [2]   Trade name for a labradorite from Finland that shows the spectral colors especially effectively.

Madagascar moonstone   Trade name for an almost transparent oligoclase moonstone (Plagioclase) from Madagascar with a strong blue schiller.

Aventurine Feldspar [5-8] Also called Sunstone   Feldspar Group

Color: Orange, red-brown, sparkling	sodium calcium aluminum silicate
Color of streak: White	Transparency: Translucent, opaque
Mohs' hardness: 6–6½	Refractive index: 1.525–1.548
Density: 2.62–2.65	Double refraction: +0.010
Cleavage: Perfect	Dispersion: None
Fracture: Grainy, splintery, brittle	Pleochroism: Weak or absent
Crystal system: (Triclinic), rare, solid aggregates	Absorption: Not diagnostic
Chemical composition: (Ca,Na)(Al,Si)$_2$Si$_2$O$_8$	Fluorescence: Dark brown-red

A metallic shiny plagioclase feldspar; named after a type of glass that was discovered by chance (Italian—*a ventura*) around 1700. Typically, it has a red, more rarely a green or blue, glitter which is caused by light reflections from tiny hematite or goethite platelets. Deposits found in India, Canada, Madagascar, Norway, Russia (Siberia), and the United States. Used with flat surfaces or en cabochon.

Possibilities for Confusion   With aventurine quartz (page 122) and the artificial glass "goldstone."

---

1 Labradorite, rough, Canada	5 Aventurine feldspar, 2 rough pieces
2 Spectrolite, rough, Finland	6 Aventurine feldspar, 4 cabochons
3 Labradorite, faceted, 4.08ct, U.S.	7 Aventurine feldspar, faceted
4 Labradorite, 13 cabochons	8 Aventurine feldspar, cabochon
The illustrations are 10 percent smaller than the originals.	

Rhodochrosite [1–5] Also called Manganesespar, Raspberryspar

Color: Rose-red to yellowish, striped
Color of streak: White
Mohs' hardness: 4
Density: 3.45–3.70
Cleavage: Perfect
Fracture: Uneven, conchoidal
Crystal system: (Trigonal) rhombohedra, usually compact aggregate
Chemical composition: $MnCO_3$ manganese carbonate
Transparency: Transparent to opaque
Refractive index: 1.600–1.820
Double refraction: –0.208 to –0.220
Dispersion: 0.015 (0.010–0.020)
Pleochroism: Absent in aggregate
Absorption spectrum: 551, 449, 415
Fluorescence: Weak; red

Rhodochrosite (Greek—rose colored) has been on the market only since about 1940. Transparent crystals are very rare. The aggregates are light-dark stripes with zigzag bands. Raspberry red and pink are the most common colors. It has a vitreous luster; on cleavage faces, there is a pearly luster. The most important deposits are in Argentina, among others those near San Luis, 144 miles (230 km) east of Mendoza. The rhodochrosite has formed as stalagmites in the silver mines of the Incas since they were abandoned in the 13th century. Other deposits are in Chile, Mexico, Peru, South Africa, and the United States.

Usually used in larger pieces, as then the marking is most distinct, for ornamental objects as well as for cabochons and necklaces. The transparent stones are usually faceted and in demand by collectors. The largest faceted rodochrosite with 59.65ct is from South Africa (private collection).

Possibilities for Confusion With fire opal (page 152), rhodonite (see below), tugtupite (page 204), tourmaline (page 110), and bustamite (page 214).

Rhodonite [6-12] Also called Manganese Gravel

Color: Dark red, flesh red, black dendritic inclusions
Color of streak: White
Mohs' hardness: 5½–6½
Density: 3.40–3.74
Cleavage: Perfect
Fracture: Uneven, conchoidal, tough
Crystal system: (Triclinic), platy, columnar, usually compact aggregate
Chemical composition: (Mn, Fe, Mg, Ca)SiO_3 manganese silicate
Transparency: Transparent to opaque
Refractive index: 1.716–1.752
Double refraction: +0.010 to +0.014
Dispersion: None
Pleochroism: Definite; yellow-red, rose-red, red-yellow in transparent
Absorption spectrum: 548, 503, 455, (412), (408)
Fluorescence: None

Rhodonite (Greek—rose), in addition to its red color, usually has black dendritic inclusions of manganese oxide. Transparent varieties are very rare. It has a vitreous luster; on cleavage faces, there is a pearly luster. Deposits are found in Australia (New South Wales), Finland, Japan, Canada, Madagascar, Mexico, Russia (the Urals), Sweden, South Africa, Tanzania, and New Jersey. It is cut with a table or en cabochon for necklaces, ornamental objects, and even wall tiles (e.g., subway in Moscow). Transparent stones are faceted with step or brilliant cut.

Possibilities for Confusion With rhodochrosite (see above), thulite (page 160), transparent varieties with hessonite (page 106), pyroxmangite (deep red manganese silicate with slight brown tint), spessartite (page 104), spinel (page 100), tourmaline (page 110), and bustamite (page 214).

Fowlerite Rhodonite variety with brownish or yellowish tint.

1 Rhodochrosite, baroque necklace	7 Rhodonite, bead necklace
2 Rhodochrosite, 3 partly polished pieces	8 Rhodonite, flat cut
3 Rhodochrosite, crystals	9 Rhodonite, unicolored cabochon
4 Rhodochrosite, 4 cabochons	10 Rhodonite, 5 transparent stones
5 Rhodochrosite, 3 cabochons	11 Rhodonite, rough, partly polished
6 Rhodonite, high cabochon	12 Rhodonite, 5 cabochons

The illustrations are 40 percent smaller than the originals.

Turquoise

Color: Sky-blue, blue-green, apple-green
Color of streak: White
Mohs' hardness: 5–6
Density: 2.31–2.84
Cleavage: None
Fracture: Conchoidal, uneven
Crystal system: (Triclinic) seldom; grape-
shaped aggregates
Chemical composition: $CuAl_6$
$(PO_4)_4(OH)_8 \cdot 4H_2O$; a copper containing

basic aluminum phosphate
Transparency: Translucent, opaque
Refractive index: 1.610–1.650
Double refraction: +0.040
Dispersion: None
Pleochroism: Absent
Absorption spectrum: (460), 432, (422)
Fluorescence: Weak; green-yellow, light
blue

The name turquoise means "Turkish stone" because the trade route that brought it to Europe used to come via Turkey. Pure blue color is rare; mostly turquoise is interspersed with brown, dark gray, or black veins of other minerals or the host rock. Such stones are called turquoise matrix. It can also be intergrown with malachite (page 176) and chrysocolla (page 200). It has a waxy luster or mat. Most of the so-called turquoise found in the United States contains Fe (substituting for Al) and is thus really a mixture with chalcosiderite. Iron imparts a greenish color.

The popular sky-blue color changes at 482 degrees F (250 degrees C) into a dull green (be careful when soldering). A negative change in color can also be brought about by the influence of light, perspiration, oils, cosmetics, and household detergents, as well as loss of natural water content. Turquoise rings should be removed before hands are washed.

Occurs in dense form, filling in fissures, as grape-like masses or nodules. Thickness of veins up to 0.8 in (20 mm). The best qualities are found in northeast Iran near Nishapur. Additional deposits are found in Afghanistan, Argentina, Australia, Brazil, China, Israel, Mexico, Tanzania, and the United States.

The deposits in Sinai, Egypt, were already worked out by 2000 B.C. In the early Victorian period, sky-blue turquoise was most popular. Today, it is used en cabochon, for brooches, necklaces, and bracelets, as well as ornamental objects.

Possibilities for Confusion   With amazonite (page 164), chrysocolla (page 200), hemimorphite (page 198), lazulite (page 192), odontolite (page 226), serpentine (page 202), smithsonite (page 198), and variscite (page 196).

Because the stone is so porous, turquoise is often soaked with artificial resin, which improves color and at the same time hardens the surface. The color can also be improved with oil or paraffin, Berliner blue, aniline colors, or copper salt. It is imitated by dyed chalcedony (page 126), dyed howlite (page 208), powdered turquoise pieces that are baked with a glue mixture, as well as by glass, porcelain, and plastics. Synthetic turquoise, with or without matrix, has been on the market with good qualities since about 1970.

Neolite (also called Reese turquoise)   Well-done turquoise imitations with dark matrix.

Neo turquoise   Good turquoise imitations with dark matrix.

Viennese turquoise   Turquoise imitations with good color.

1 Turquoise with matrix, 2 cabochons	8 Turquoise, 2 matrix cabochons
2 Turquoise, Chinese carving	9 Turquoise, 7 cabochons, 14.30ct
3 Turquoise, 3 cabochons, 25.89ct	10 Turquoise, baroque necklace
4 Turquoise, rough	11 Turquoise, rough, partly polished
5 Turquoise, 9 cabochons, 26.10ct	12 Turquoise, 2 cabochons, 38.53ct
6 Turquoise, 3 cabochons, octagon	13 Turquoise, 4 cabochons, 42.48ct
7 Turquoise, bead necklace	14 Turquoise, 3 cabochons

The illustrations are 40 percent smaller than the originals.

Lapis Lazuli

Color: Lazur blue, violet, greenish-blue	calcium aluminum silicate
Color of streak: Light blue	Transparency: Opaque
Mohs' hardness: 5–6	Refractive index: About 1.50
Density: 2.50–3.00	Double refraction: None
Cleavage: Indistinct	Dispersion: None
Fracture: Conchoidal, grainy	Pleochroism: Absent
Crystal system: (Cubic) rare; dense aggregates	Absorption spectrum: Not diagnostic
Chemical composition:	Fluorescence: Strong: white, also orange, copper-colored
(Na,Ca)$_8$[(SiO$_4$,S,Cl)$_2$](AlSi)$_4$]$_6$ sodium	

As lapis lazuli (Arabic and Latin—blue stone) is composed of several minerals—in addition to lazurite (25–40 percent), including augite, calcite, diopside, enstatite, mica, haüynite, hornblende, nosean, sodalite, and/or pyrite—it is considered to be not a mineral but a rock. The variation in composition also causes a wide range in the above data.

The coloring agent is sulfur. In the best-quality specimens, the color is evenly distributed, but in general it is spotty or striated. In the lapis lazuli from Chile (Chilean lapis) and from Russia, a strongly protruding whitish or gray calcite diminishes the value. Well-distributed fine pyrite is advantageous and is taken to show genuineness. Too much pyrite, on the other hand, causes a dull, greenish tint. Color can be improved through slight heating and dyeing. It has a vitreous and greasy luster.

Lapis lazuli is sensitive to strong pressure and high temperatures, hot baths, acids and alkalies. Rings should be taken off during household work!

For 6,000 years, the deposit producing the finest material has been in the West Hindu Kush mountains in Afghanistan. Lapis lazuli is present there as an irregular occurrence in limestone in difficult terrain. Russian deposits are at the southwest end of Lake Baikal. The matrix rock is white dolmitic marble. Chile's deposits are north of Santiago. Other deposits are found in Angola, Burma (Myanmar), Canada, Pakistan, California, and Colorado.

Lapis lazuli was already used in prehistoric times for jewelry. During the Middle Ages, it was also used as a pigment to produce ultramarine. Some palaces and churches have wall panels and columns inlaid with lapis lazuli. Today it is used for ring stones and necklaces, as well as sculptures, vases, and other ornamental objects.

Possibilities for Confusion   With azurite (page 174), dumortierite (page 182), dyed howlite (page 208), lazulite (page 192), sodalite (page 174), and glass imitations.

In 1954, a synthetic grainy spinel, colored with cobalt oxide, with a good lapis color, made an appearance on the market. Inclusions of thin gold pieces simulated the pyrite. Imitations with lapis lazuli pieces and powder, pressed or bound with artificial resin, are also on the market.

German lapis (also called Swiss lapis)   Misleading trade name for a dyed jasper (see page 146). It has no pyrite inclusions.

1 Lapis lazuli, bowl, Chile	8 Lapis lazuli, 3 cabochons, Afghanistan
2 Lapis lazuli, rough, Afghanistan	9 Lapis lazuli ring stone, Siberia, Russia
3 Lapis lazuli, broken crystal	10 Lapis lazuli, 7 different cuts
4 Lapis lazuli, Afghanistan, partly polished	11 Lapis lazuli, tablet cut, Chile
5 Lapis lazuli, Buddha, Afghanistan	12 Lapis lazuli, rough, Siberia, Russia
6 Lapis lazuli, bead necklace, Afghanistan	13 Lapis lazuli, rough, Afghanistan
7 Lapis lazuli, cabochon, Afghanistan	14 Lapis lazuli, rough, Chile

The illustrations are 40 percent smaller than the originals.

Sodalite [1-4]

Color: White, blue, gray
Color of streak: White
Mohs' hardness: 5½–6
Density: 2.14–2.40
Cleavage: Indistinct
Fracture: Uneven, conchoidal
Crystal system: (Cubic) rhombic
    dodecahedra
Chemical composition: $Na_8Al_6Si_6O_{24}Cl_2$

chloric sodium aluminum silicate
Transparency: Transparent to opaque
Refractive index: 1.48
Double refraction: None
Dispersion: 0.018 (0.009)
Pleochroism: Absent
Absorption spectrum: Not diagnostic
Fluorescence: Strong; orange

The name sodalite refers to its sodium content. For jewelry, only blue tones are used; sometimes they have a violet tint; frequently they are dispersed with white veins from white calcite. It has a vitreous luster; on fractures, there is a greasy luster. The low density is quite noticeable. Occurrence is usually in syenite and trachyte rocks, as well as pegmatite. Deposits are found in Brazil (Bahia), Greenland, India, Canada (Ontario), Namibia (transparent crystals), Russia (Urals), and Montana. Used as cabochon and for necklaces, especially for arts and crafts objects, also as ornamental stone.

Possibilities for Confusion   With azurite (see below), dumortierite (page 182), haüynite (page 204), lapis lazuli (page 172), and lazulite (page 192). Synthetic sodalite has been known since 1975.

Hackmanite   Pink-colored sodalite variety. It was discovered for the first time in cuttable qualities in 1991, in Quebec, Canada. The color fades in light.

Azurite [5-8] Also called Chessylite

Color: Dark blue, azure blue
Color of streak: Sky-blue
Mohs' hardness: 3½–4
Density: 3.7–3.9
Cleavage: Indistinct
Fracture: Conchoidal, uneven, brittle
Crystal system: (Monoclinic); short
    columnar, dense aggregates
Chemical composition: $Cu_3(CO_3)_2(OH)_2$

basic copper carbonate
Transparency: Transparent to opaque
Refractive index: 1.720–1.848
Double refraction: +0.108 to +0.110
Dispersion: None
Pleochroism: Definite; light blue, dark blue
Absorption spectrum: 500
Fluorescence: None

Named after its azure-blue color. It has a vitreous luster. Occurs with malachite in or near copper deposits. Exceptional dark blue—almost black—crystals have come from Tsumeb, Namibia, with similar material now coming from Morocco. Deposits are found in Australia (Queensland), Chile, Mexico, Russia (Urals), and the United States (Arizona, New Mexico). The famous deposit in Chessy near Lyon, France, seems to be worked out.

It was formerly used for azure pigment. Because of its low hardness, it is mainly used for ornamental objects. It is also cut by collectors en cabochon and even faceted.

Possibilities for Confusion   With dumortierite (page 182), haüynite (page 204), lapis lazuli (page 172), lazulite (page 192), and sodalite (see above).

Azure-malachite [8]   Ball- or kidney-shaped, striped intergrowth of azurite and malachite (page 176). Frequently cut en cabochon.

Burnite   Intergrowth of azurite and cuprite (page 206).

---

1 Sodalite, rough, partly polished
2 Sodalite, baroque necklace
3 Sodalite, 4 flat cut pieces
4 Sodalite, 2 cabochons
The illustrations are 20 percent smaller than the originals.

5 Azurite, crystals
6 Azurite, 5 differently cut stones
7 Azurite, part of crystal
8 Azure-malachite, rough

Malachite

Color: Light- to black-green, banded	basic copper carbonate
Color of streak: Light green	Transparency: Translucent, opaque
Mohs' hardness: 3½–4	Refractive index: 1.655–1.909
Density: 3.25–4.10	Double refraction: −0.254
Cleavage: Perfect	Dispersion: None
Fracture: Splintery, scaly	Pleochroism: Absent
Crystal system: (Monoclinic) small, long prismatic; usually aggregate	Absorption spectrum: Not diagnostic
	Fluorescence: None
Chemical composition: $Cu_2(CO_3)(OH)_2$	

The name is probably derived from the green color (Greek—*malache* = mallow), perhaps from its low hardness (Greek—*malakos* = soft). In fracture or when cut, aggregates show a banding of light and dark layers with concentric rings, straight stripes, or other figurative shapes caused by its shell-like formation. Large mono-colored pieces are rare. In thin plates, it is translucent, otherwise opaque. The coloring agent is copper. Crystals are rare, mostly dense, fibrous fine-crystalline aggregates. As rough stone, it has a weak vitreous or mat luster; on fresh fractures and when polished, it has a silky luster. Malachite is sensitive to heat, acids, ammonia, and hot waters.

Occurs in rounded nodules, grape shapes, cone shapes, or stalactitic and, rarely, encrusted slabs. Formed from copper-containing solutions in or near copper ore deposits. The most important deposits used to be in the Urals near Yekaterinburg (Sverdlovsk). The quarries delivered blocks of over 20 tons in weight. From there, the Russian tsars obtained the malachite for decorating their palaces, paneling the walls, and for beautiful inlaid works.

Today Shaba (Katanga) in Zaire near Zambia is the most important malachite producer. Other deposits are in Australia (Queensland, New South Wales), Chile, Namibia, Zimbabwe, and Arizona.

Malachite was popular with the ancient Egyptians, Greeks, and Romans for jewelry, amulets, and as a powder for eye shadow. It is used as pigment for mountain green.

Although it is not very hard and not very resistant, malachite is nowadays popular for jewelry and ornaments. Used en cabochon and as slightly rounded tablet stones for necklaces and especially for objets d'art, such as plates, boxes, ashtrays, and sculptures. The cutter must work the malachite so as to show the decorative marking to its best advantage. Concentric eye-like rings are most popular (called malachite peacock's eye). Because of its low hardness, malachite is easily scratched and sometimes becomes dull. The surface can be hardened with artificial resin.

Possibilities for Confusion Unlikely for larger pieces because of its striped formation; small, unbanded stones, on the other hand, can be confused with opaque green gemstones.

Azure-malachite Intergrowth of azurite and malachite (see page 174).

Eilat stone Intergrowth of malachite with turquoise and chrysocolla (see page 200).

---

1 Malachite, rough, partly polished	4 Malachite, cabochon, Zimbabwe
2 Malachite, bead necklace	5 Malachite, 7 various samples
3 Malachite, 2 cabochons	6 Malachite, rough

The illustrations are 40 percent smaller than the originals.

Lesser-Known Gemstones

Those gemstones that are not commonly known but which are becoming increasingly more popular belong to this group. Compare also with page 69.

Andalusite [2-3]

Color: Yellow-green, green, brownish-red
Color of streak: White
Mohs' hardness: 7½
Density: 3.05–3.20
Cleavage: Good
Fracture: Uneven, brittle
Crystal system: (Orthorhombic), thick-columnar
Chemical composition: Al_2SiO_5 aluminum silicate

Transparency: Transparent to opaque
Refractive index: 1.627–1.649
Double refraction: –0.007 to –0.013
Dispersion: 0.016 (0.009)
Pleochroism: Strong; yellow, olive, red-brown to dark red
Absorption spectrum: 553, <u>550</u>, 547, (525), (518), (495), <u>455</u>, 447, <u>436</u>
Fluorescence: Weak; green, yellow-green

Transparent andalusite (named after Andalusia in Spain) of gemstone quality is rare. The luster is vitreous or mat. When cutting, the strong pleochroism has to be taken into consideration. Occurs in schists, gneisses, and in placers. Deposits are found in Australia, Brazil, Canada, Russia, Spain (Andalusia), Sri Lanka, and the United States. Used in table and brilliant cut.

Possibilities for Confusion   With chrysoberyl (page 98), smoky quartz (page 116), sinhalite (page 186), sphene (page 194), tourmaline (page 110), and idocrase (page 186).

Chiastolite [1] (Also called Cross Stone)   Opaque variety of andalusite; white, gray, yellowish; Mohs' hardness is 5–5½. Occurs in long prisms that show a dark cross in cross section, when viewed perpendicular to the prism axes; caused by carbonaceous inclusions. Deposits are found in Algeria, South Australia, Bolivia, Chile, France (Brittany), Russia (Kola, Siberia), Spain (Galicia), and California. Worn as an amulet; now of value only to the collector. Cut en cabochon, flat.

Possibilities for Confusion   Do not exist due to its distinct marking.

Viridine   Dark-green andalusite variety, containing iron and manganese.

Euclase [4-6]

Color: Colorless, sea-green, light blue, dark blue
Color of streak: White
Mohs' hardness: 7½
Density: 3.10
Cleavage: Perfect
Fracture: Conchoidal, brittle
Crystal system: Monoclinic; prisms
Chemical composition: $BeAlSiO_4(OH)$ basic beryllium aluminum silicate

Transparency: Transparent
Refractive index: 1.650–1.677
Double refraction: +0.019 to +0.025
Dispersion: 0.016 (0.009)
Pleochroism: Very weak; white-green, yellow-green, blue-green
Absorption spectrum: <u>706</u>, <u>704</u>, <u>650</u>, <u>639</u>, 468, 455
Fluorescence: Weak or none

Euclase (Greek—breaks well) is difficult to cut and polish because of its perfect cleavage. It has a bright, vitreous luster. Occurs in pegmatites, in placers, and in druses. Deposits are found in Brazil (Minas Gerais), Russia (Urals), Zimbabwe, Tanzania, and Zaire. Used with a step cut.

Possibilities for Confusion   With aquamarine (page 94), precious beryl (page 96), hiddenite (page 114), and sapphire (page 86). With radiation, colorless euclase can be changed to blue.

1 Chiastolite, 4 pieces, partly polished	4 Euclase, 2 colorless faceted stones
2 Andalusite, 2 broken crystals	5 Euclase, light blue, faceted
3 Andalusite, 4 faceted stones	6 Euclase, crystal in host rock

Hambergite [1, 2]

Color: Colorless, gray-white, yellow-white
Color of streak: White
Mohs' hardness: 7½
Density: 2.35
Cleavage: Perfect
Fracture: Conchoidal, brittle
Crystal system: Orthorhombic; prisms
Chemical composition: $Be_2BO_3(OH)$

Transparency: Transparent, translucent
Refractive index: 1.553–1.628
Double refraction: +0.072
Dispersion: (0.015 (0.009–0.010)
Pleochroism: Absent
Absorption spectrum: Cannot be evaluated
Fluorescence: Usually none, sometimes orange

Hambergite (after a Swedish mineralogist) has the same vitreous luster as glass when cut. The strong double refraction can be recognized through the table along the lower facet edges.

Possibilities for Confusion   With rock crystal (page 116), danburite (page 182), euclase (page 178), leuko garnet (page 106), and zircon (page 108).

Iolite [3-6]   Also called Cordierite and Dichroite

Color: Blue, violet, brownish
Color of streak: White
Mohs' hardness: 7–7½
Density: 2.58–2.66
Cleavage: Good
Fracture: Conchoidal, uneven, brittle
Crystal system: Orthorhombic; short prisms
Chemical composition: $Mg_2Al_4Si_5O_{18}$ magnesium aluminum silicate

Transparency: Transparent, translucent
Refractive index: 1.542–1.578
Double refraction: –0.008 to –0.012
Dispersion: 0.017 (0.009)
Pleochroism: Very strong; yellow, dark blue-violet, pale blue
Absorption spectrum: 645, 593, 585, 535, 492, 456, 436, 426
Fluorescence: None

The color of iolite (Greek—violet) is usually blue. Inclusions of hematite and goethite sometimes cause reddish sheen or aventurescence. It has a greasy vitreous luster. Deposits found in Burma (Myanmar), Brazil, India, Madagascar, Sri Lanka, and the United States.

Possibilities for Confusion   With benitoite (page 184), kyanite (page 196), sapphire (page 86), and tanzanite (page 160). Glass imitations are known.

Water sapphire   Misleading trade name for blue iolite.

Phenakite [7, 8]

Color: Colorless, wine-yellow, pink
Color of streak: White
Mohs' hardness: 7½–8
Density: 2.95–2.97
Cleavage: Good
Fracture: Conchoidal
Crystal system: (Trigonal) short columnar
Chemical composition: Be_2SiO_4 beryllium silicate

Transparency: Transparent
Refractive index: 1.650–1.670
Double refraction: +0.016
Dispersion: 0.015 (0.009)
Pleochroism: Definite; colorless, orange-yellow
Absorption spectrum: Not diagnostic
Fluorescence: Pale greenish, blue

Phenakite (Greek—deceiver) is mostly water-clear. It has a strong vitreous luster; when polished, a greasy luster. Colored stones can fade. Deposits are found in Brazil, Mexico, Namibia, Norway, Zimbabwe, Sri Lanka, Tanzania, and the United States.

Possibilities for Confusion   With rock crystal (page 116), beryllonite (page 190), cerussite (page 200), danburite (page 182), precious beryl (page 96), and topaz (page 102).

1 Hambergite, 3 faceted stones	5 Iolite, 2 cut cubes
2 Hambergite, 2 broken crystals	6 Iolite, 3 ovals
3 Iolite, 6 faceted stones	7 Phenakite, 3 rough pieces
4 Iolite, 2 rough pieces	8 Phenakite, 2 faceted stones

Dumortierite [1, 2]

Color: Dark blue, violet-blue, red-brown, colorless
Color of streak: White
Mohs' hardness: 7–8½
Density: 3.26–3.41
Cleavage: Good
Fracture: Conchoidal
Crystal system: (Orthorhombic) very rare; fibrous or radial aggregates
Chemical composition: $Al_7(BO_3)(SiO_4)_3O_3$
aluminum borate silicate
Transparency: Transparent to opaque
Refractive index: 1.678–1.689
Double refraction: –0.015 to –0.037
Dispersion: None
Pleochroism: Strong; black, red-brown, brown
Absorption: Not diagnostic
Fluorescence: Weak; blue, blue-white, violet

Aggregates of dumortierite, mainly cut, have a Mohs' hardness of 7; crystals have 8½. Named after a French palaeontologist. Deposits are found in Brazil, France, India, Canada, Madagascar, Mozambique, Namibia, Sri Lanka, and the United States.

Possibilities for Confusion   With azurite (page 174), blue quartz (page 122), lapis lazuli (page 172), and sodalite (page 174).

Dumortierite quartz   Quartz which is intergrown with dumortierite [1].

Danburite [3, 4]

Color: Colorless, wine-yellow, brown, pink
Color of streak: White
Mohs' hardness: 7–7½
Density: 2.97–3.03
Cleavage: Imperfect
Fracture: Uneven, conchoidal
Crystal system: Orthorhombic; prismatic
Chemical composition: $CaB_2(SiO_4)_2$
calcium boric silicate
Transparency: Transparent
Refractive index: 1.630-1.636
Double refraction: –0.006 to –0.008
Dispersion: 0.017 (0.009)
Pleochroism: Weak; light yellow
Absorption spectrum: 590, 586, 585, 584, 583, 582, 580, 578, 573, 571, 658, 566, 564
Fluorescence: Sky-blue

Named after the first place of discovery in Danbury, Connecticut. It has a greasy, vitreous luster. It can be well faceted because of its hardness and little cleavage. Deposits are found in Burma (Myanmar), Japan, Madagascar, Mexico, Russia, and Connecticut.

Possibilities for Confusion   With citrine (page 120), hambergite (page 180), phenakite (page 180), and topaz (page 102).

Axinite (group) [5, 6]

Color: Brown, violet, blue
Color of streak: White
Mohs' hardness: 6½–7
Density: 3.26–3.36
Cleavage: Good
Fracture: Conchoidal, brittle
Crystal system: Triclinic; platy crystals
Chemical composition:
   $(Ca,Fe,Mn,Mg)_3Al_2BSi_4O_{15}(OH)$
   calcium aluminum borate silicate
Transparency: Translucent, transparent
Refractive index: 1.656–1.704
Double refraction: –0.010 to –0.012
Dispersion: 0.018–0.020 (0.011)
Pleochroism: Strong; olive-green, red-brown, violet-blue
Absorption spectrum: 532, 512, 492, 466, 444, 415
Fluorescence: Red, orange

The axinite group includes ferroaxinite, manganaxinite, magnesioaxenite, and tinzenite. Named (Greek—axe) because of its sharp-edged crystals. It has a strong, vitreous luster. As it is pyro- and piezo-electric, azinite attracts dust and must be cleaned frequently. Deposits are found in Brazil, England (Cornwall), France (Pyrenees, Dep. Isère), Mexico (Baja California), Russia (Urals), Sri Lanka, Tanzania, and California.

Possibilities for Confusion   With andalusite (page 178), barite (page 206), smoky quartz (page 116), and sphene (page 194).

1 Dumortierite, quartz, California	4 Danburite, 3 broken crystals
2 Dumortierite, 2 cabochons	5 Axinite, 5 different cuts
3 Danburite, 9 different cuts	6 Axinite, rough

Benitoite [1, 2]

Color: Blue, purple, pink, colorless	barium titanium silicate
Color of streak: White	Transparency: Transparent
Mohs' hardness: 6–6½	Refractive index: 1.757–1.804
Density: 3.64–3.68	Double refraction: +0.047
Cleavage: Indistinct	Dispersion: 0.046 (0.026)
Fracture: Conchoidal, brittle	Pleochroism: Very strong; colorless, blue
Crystal system: (Hexagonal) bipyramidal	Absorption spectrum: Not diagnostic
Chemical composition: BaTiSi$_3$O$_9$	Fluorescence: Strong; blue

Named after the first place of discovery in San Benito County, California. Only small crystals are found of gemstone quality. It has a vitreous to adamantine luster. Deposits found only in California.

Possibilities for Confusion   With iolite (page 180), kyanite (page 196), sapphire (page 86), spinel (page 100), tanzanite (page 160), tourmaline (page 110), and zircon (page 108).

Cassiterite [3–5] Also called Tin Stone

Color: Various browns, colorless	Transparency: Transparent to opaque
Color of streak: White to light yellow	Refractive index: 1.997–2.098
Mohs' hardness: 6–7	Double refraction: +0.096 to +0.098
Density: 6.7–7.1	Dispersion: 0.071 (0.035)
Cleavage: Indistinct	Pleochroism: Weak to strong; green-yellow,
Fracture: Conchoidal, brittle	brown, red-brown
Crystal system: Tetragonal; short	Absorption spectrum: Not diagnostic
columnar crystals	Fluorescence: None
Chemical composition: SnO$_2$ tin oxide	

Cassiterite (Greek—tin) has anadamantine luster. Deposits are found in Australia (New South Wales), Bolivia, England (Cornwall), Malaysia, Mexico, Namibia, Spain, and California.

Possibilities for Confusion   With diamond (page 70), smoky quartz (page 116), scheelite (page 196), sinhalite (page 186), sphene (page 194), idocrase (page 186), and zircon (page 108).

Epidote [6–8] Also called Pistacite

Color: Pistachio green	calcium aluminum iron silicate
Color of streak: Gray	Transparency: Transparent to opaque
Mohs' hardness: 6–7	Refractive index: 1.729–1.768
Density: 3.3–3.5	Double refraction: +0.015 to –0.049
Cleavage: Perfect	Dispersion: 0.030 (0.012–0.027)
Fracture: Conchoidal, splintery	Pleochroism: Strong; green, brown, yellow
Crystal system: Monoclinic; prisms	Absorption spectrum: 475, 455, 435
Chemical composition:	Fluorescence: None
Ca$_2$(Fe,Al)$_3$(SiO$_4$)$_3$(OH)	

Named (Greek—addition) after the numerous crystal faces It has a bright, vitreous luster. Deposits are found in Brazil, Kenya, Mexico, Mozambique, Norway, Austria (Untersulzbach Valley/Salzburg), Sri Lanka, and California.

Possibilities for Confusion   With diopside (page 190), dravite (page 110), and idocrase (page 186).

Clinozoisite   Epidote mineral, low on iron. (Compare with page 212, picture on page 213, no. 17.)

Piemontite (piedmontite)   Opaque, red, manganese-containing epidote mineral.

Tawmawite   Dark green, chrome-containing epidote variety from Burma (Myanmar).

1 Benitoite, 2 crystal formations
2 Benitoite, 8 faceted stones
3 Cassiterite, crystals, Cornwall, England
4 Cassiterite, 3 light brown stones, Malaysia
5 Cassiterite, crystal, Cornwall
6 Epidote, 3 faceted stones
7 Epidote, 2 broken crystals
8 Epidote, twinned aggregate
The illustrations are 30 percent larger than the originals.

Idocrase [1–4] Also called Vesuvianite

Color: Olive-green, yellow-brown, pale blue
Color of streak: White
Mohs' hardness: 6½
Density: 3.32–3.47
Cleavage: Indistinct
Fracture: Uneven, splintery
Crystal system: Tetragonal; thick columnar crystals
Chemical composition:
$Ca_{10}Mg_2Al_4(SiO_4)_5(Si_2O_7)_2(OH)_4$ complicated calcium aluminum silicate
Transparency: Transparent, translucent
Refractive index: 1.700–1.723
Double refraction: +0.002 to –0.012
Dispersion: 0.019–0.025 (0.014)
Pleochroism: Weak; present color lighter and darker
Absorption spectrum: Green: (528), 461; Brown: 591, 588, 584, 582, 577, 574
Fluorescence: None

Due to wide variations in composition, there is a range of physical properties displayed. The luster is greasy. Deposits are found in Brazil, Mexico, Kenya, Russia, Switzerland, Sri Lanka, and the United States.

Possibilities for Confusion   With demantoid (page 106), diopside (page 190), epidote (page 184), zircon (page 108), peridot (page 158), and sinhalite (see below).

Californite   Translucent green variety; falsely called California jade.

Cyprine   Sky-blue variety from Norway.

Sinhalite [5, 6]

Color: Yellow-brown, green-brown
Color of streak: White
Mohs' hardness: 6½–7
Density: 3.46–3.50
Cleavage: None
Fracture: Conchoidal
Crystal system: (Orthorhombic) very rare; grains
Chemical composition: $MgAlBO_4$ magnesium aluminum iron borate
Transparency: Transparent, translucent
Refractive index: 1.665–1.712
Double refraction: –0.036 to –0.042
Dispersion: 0.018 (0.010)
Pleochroism: Definite; green, light brown, dark brown
Absorption spectrum: 526, 492, 475, 463, 452
Fluorescence: None

First recognized as an individual mineral in 1952 in Sri Lanka, then Ceylon (Sanskrit—*Sinhala*). It has a vitreous luster. Deposits found in Burma (Myanmar), Russia, Sri Lanka, and Tanzania.

Possibilities for Confusion   With chrysoberyl (page 98), peridot (page 158), tourmaline (page 110), idocrase (see above), and zircon (page 108).

Kornerupine [7, 8] Also called Prismatine

Color: Green, green-brown
Color of streak: White
Mohs' hardness: 6½–7
Density: 3.27–3.45
Cleavage: Good
Fracture: Conchoidal
Crystal system: Orthorhombic; long prisms
Chemical composition:
$Mg_3Al_6(Si,Al,B)_5O_{21}(OH)$ magnesium aluminum borate silicate
Transparency: Transparent, translucent
Refractive index: 1.660–1.699
Double refraction: –0.012 to –0.017
Dispersion: 0.018 (0.010)
Pleochroism: Strong; green, yellow, reddish-brown
Absorption spectrum: 540, 503, 463, 446, 430
Fluorescence: Usually none; green k. from Kenya: yellow

Named after a Danish geologist and explorer of Greenland. It has a vitreous luster. Deposits are found in Burma (Myanmar), Canada (Quebec), Kenya, Madagascar, Sri Lanka, Tanzania, and South Africa.

Possibilities for Confusion   With enstatite (page 192), epidote (page 184), and tourmaline (page 110).

1 Idocrase, crystal
2 Idocrase, 3 faceted stones, 6.25ct
3 Idocrase, cabochon, 4.19ct
4 Idocrase, 4 faceted stones
5 Sinhalite, rough
6 Sinhalite, 2 faceted stones
7 Kornerupine, 3 faceted stones
8 Kornerupine, aggregate, Sri Lanka
The illustrations are 20 percent larger than the originals.

Prehnite [1–3]

Color: Yellow-green, brown-yellow
Color of streak: White
Mohs' hardness: 6–6½
Density: 2.82–2.94
Cleavage: Good
Fracture: Uneven
Crystal system: Orthorhombic; columnar, tabular crystals and aggregates
Chemical composition:

$Ca_2Al_2Si_3O_{10}(OH)_2$
basic calcium aluminum silicate
Transparency: Transparent, translucent
Refractive index: 1.611–1.669
Double refraction: +0.021 to +0.039
Dispersion: None
Pleochroism: Absent
Absorption spectrum: 438
Fluorescence: None

Named after Dutch colonel. It has a vitreous to mother-of-pearl luster. Prehnite cat's-eye occurs. Deposits are found in Australia, China, Scotland, South Africa, and the United States.

Possibilities for Confusion With apatite (page 194), brazilianite (page 190), chrysoprase (page 128), jade (page 154), peridot (page 158), periclase (page 206), and serpentine (page 202).

Petalite [4, 5]

Color: Colorless, pink, yellowish
Color of streak: White
Mohs' hardness: 6–6½
Density: 2.40
Cleavage: Perfect
Fracture: Conchoidal, brittle
Crystal system: (Monoclinic) thick tabular, columnar and aggregates

Chemical composition: $LiAlSi_4O_{10}$ lithium aluminum silicate
Refractive index: 1.502–1.519
Double refraction: +0.012 to +0.017
Dispersion: Unknown
Pleochroism: Absent
Absorption spectrum: (454)
Fluorescence: Weak; orange

Named (Greek—leaf) because of its perfect cleavage. It has a vitreous luster; pearly luster on cleavage planes. Crystals are rare; mostly coarse aggregates occur; and petalite cat's-eye is known. Deposits are found in Western Australia, Brazil (Minas Gerais), Italy (Elba), Namibia, Sweden, Zimbabwe, and the United States.

Possibilities for Confusion With other colorless gemstones and glass.

Scapolite [6–8] Also called Wernerite

Color: Yellow, pink, violet, colorless
Color of streak: White
Mohs' hardness: 5½–6
Density: 2.57–2.74
Cleavage: Good
Fracture: Conchoidal, brittle
Crystal system: Tetragonal; columnar
Chemical composition: $Na_4Al_3Si_9O_{24}Cl$ to $Ca_4Al_6Si_6O_{24}(CO_3,SO_4)$ sodium calcium aluminum silicate

Transparency: Transparent, translucent
Refractive index: 1.540–1.579
Double refraction: −0.006 to −0.037
Dispersion: 0.017
Pleochroism: Definite; yellow s.: colorless, yellow
Absorption spectrum: 663, 652
Fluorescence: Pink: orange, pink; Yellow: violet, blue-red

Scapolite is named (Greek—stick) after its crystal habit. It has a vitreous luster. Pink and violet scapolite cat's-eyes are known. Deposits are found in Burma (Myanmar), Brazil, Canada, Madagascar, and Tanzania.

Possibilities for Confusion With amblygonite (page 192), chrysoberyl (page 98), golden beryl (page 96), rose quartz (page 122), sphene (page 194), and tourmaline (page 110).

Petschite A violet scapolite variety from Tanzania, discovered in 1975.

1 Prehnite, 2 cabochons, 31.91ct
2 Prehnite, 2 faceted stones, Australia
3 Prehnite, with apophyllite crystals
4 Petalite, rough

5 Petalite, 3 faceted stones
6 Scapolite, 5 faceted stones
7 Scapolite cat's-eye, 5 stones
8 Scapolite, 4 crystal pieces

Diopside [1–3]

Color: Green, yellow, colorless, brown, black
Color of streak: White
Mohs' hardness: 5–6
Density: 3.22–3.38
Cleavage: Good
Fracture: Uneven, rough
Crystal system: Monoclinic; columnar crystals
Chemical composition: $CaMgSi_2O_6$ calcium magnesium silicate

Transparency: Transparent, translucent
Refractive index: 1.664–1.730
Double refraction: +0.024 to +0.031
Dispersion: 0.017–0.020 (0.012)
Pleochroism: Weak; yellow-green, dark green
Absorption spectrum: (505), (493), (446); Chrome diopside: (690), (670), (655), (635), 508, 505, 490
Fluorescence: Violet, orange, yellow, green

Named (Greek—double appearance) because of its crystal shape. Deposits are found in Burma (Myanmar), Finland, India, Madagascar, Austria, Sri Lanka, South Africa, and the United States. Star diopside [3] and diopside cat's-eye are known.

Possibilities for Confusion With hiddenite (page 114), moldavite (page 220), peridot (page 158), emerald (page 90), and idocrase (page 186).

Chrome diopside Diopside variety with strong emerald green color.

Violane Coarse, violet-blue diopside variety; translucent to opaque from Piedmont, Italy. Used for ornamental objects.

Beryllonite [4]

Color: Colorless, white, weak yellow
Color of streak: White
Mohs' hardness: 5½–6
Density: 2.80–2.87
Cleavage: Perfect
Fracture: Conchoidal, brittle
Crystal system: Monoclinic; short prisms
Chemical composition: $NaBePO_4$ sodium beryllium phosphate

Transparency: Transparent
Refractive index: 1.552–1.561
Double refraction: −0.009
Dispersion: 0.010 (0.007)
Pleochroism: Absent
Absorption spectrum: Cannot be evaluated
Fluorescence: None

So named because of its beryllium content. It has a vitreous luster; on fracture planes, pearly luster. Deposits are found in Brazil, Finland, Zimbabwe, and Maine.

Possibilities for Confusion With other colorless gemstones and glass.

Brazilianite [5, 6]

Color: Yellow, green-yellow
Color of streak: White
Mohs' hardness: 5½
Density: 2.98–2.99
Cleavage: Good
Fracture: Small conchoidal, brittle
Crystals: Monoclinic; short prisms
Chemical composition: $NaAl_3(PO_4)_2(OH)_4$ sodium aluminum phosphate

Transparency: Transparent, translucent
Refractive index: 1.602–1.623
Double refraction: +0.019 to +0.021
Dispersion: 0.014 (0.008)
Pleochroism: Very weak
Absorption spectrum: Not diagnostic
Fluorescence: None

Named after Brazil, the first country of discovery (1944). Deposits are found in Brazil (Minas Gerais) and the United States (New Hampshire).

Possibilities for Confusion With amblygonite (page 192), apatite (page 194), chrysoberyl (page 98), precious beryl (page 96), and topaz (page 102).

---

1 Diopside, 10 different cuts
2 Diopside, 2 broken crystals
3 Diopside, 4-rayed asterism
The illustrations are 15 percent larger than the originals.

4 Beryllonite, 3 faceted stones
5 Brazilianite, rough
6 Brazilianite, 5 faceted stones

1

2

2

2

3

4

5

6

Amblygonite [1, 2]

Color: Golden-yellow to colorless, purple	basic lithium aluminum phosphorate
Color of streak: White	Transparency: Transparent, translucent
Mohs' hardness: 6	Refractive index: 1.578–1.646
Density: 3.01–3.11	Double refraction: +0.024 to +0.030
Cleavage: Perfect	Dispersion: 0.014–0.015 (0.008)
Fracture: Uneven, brittle	Pleochroism: Absent
Crystal system: (Triclinic) prismatic	Absorption spectrum: Not diagnostic
Chemical composition: (Li,Na)Al(PO$_4$)(F,OH)	Fluorescence: Very weak; green

So named (Greek—crooked angled) because of its crystal shape. Vitreous luster, pearly luster on cleavage planes. Deposits: Burma, Brazil (Minas Gerais, Sao Paulo), Sweden, and California. A purple variety from Namibia is known.

Possibilities for Confusion   With apatite (page 194), brazilianite (page 190), citrine (page 120), golden beryl (page 96), hiddenite (page 114), and scapolite (page 188).

Enstatite [3, 4]

Color: Brown-green, green, colorless, yellowish	Transparency: Transparent to opaque
Color of streak: White	Refractive index: 1.650–1.680
Mohs' hardness: 5½	Double refraction: +0.009 to + 0.012
Density: 3.20–3.30	Dispersion: Little to none (0.010)
Cleavage: Good	Pleochroism: Definite; green, yellow-green
Fracture: Scaly	Absorption spectrum: 547, 509, 505, 502, 483, 459, 449; Chrome e.: 688, 669, 506
Crystal system: Orthorhombic; prismatic	
Chemical composition: Mg$_2$Si$_2$O$_6$ magnesium silicate	Fluorescence: None

So named (Greek—resistor) because it does not melt easily. It has a vitreous luster. Deposits are found in Burma (Myanmar), Brazil, India, Kenya, Mexico, Sri Lanka, South Africa, Tanzania, and the United States. Greenish-gray enstatite cat's-eye and star enstatite are known.

Possibilities for Confusion   With andalusite (page 178), kornerupine (page 186), sphalerite (page 200), idocrase (page 186), and zircon (page 108).

Bronzite   Green-brown enstatite with metallic luster; rich in iron.

Lazulite [5, 6] Also called Bluespar

Color: Dark blue to blue-white, green-blue	basic magnesium aluminum phosphate
Color of streak: White	Transparency: Transparent to opaque
Mohs' hardness: 5–6	Refractive index: 1.612–1.646
Density: 3.04–3.14	Double refraction: –0.031 to –0.036
Cleavage: Indistinct	Dispersion: Unknown
Fracture: Uneven, splintery, brittle	Pleochroism: Strong; colorless, dark blue
Crystal system: Monoclinic; pointed pyramids	Absorption spectrum: Not diagnostic
	Fluorescence: None
Chemical composition: MgAl$_2$(PO$_4$)$_2$(OH)$_2$	

Named (Persian and Greek—blue stone) because of its color. It has a vitreous luster. Deposits are found in Angola, Bolivia, Brazil (Minas Gerais), India, Madagascar, Austria (Salzburg), Sweden, and North Carolina.

Possibilities for Confusion   With azurite (page 174), iolite (page 180), indicolite (page 110), lapis lazuli (page 172), sodalite (page 174), topaz (page 102), and turquoise (page 170).

1 Amblygonite, 2 rough pieces
2 Amblygonite, 6 different cuts
3 Enstatite, 2 faceted stones and Star enstatite
4 Enstatite, 3 rough pieces
5 Lazulite, 9 different cuts
6 Lazulite, rough
The illustrations are 20 percent larger than the originals.

Dioptase [1, 2]

Color: Emerald green, blue-green
Color of streak: Greenish
Mohs' hardness: 5
Density: 3.28–3.35
Cleavage: Perfect
Fracture: Conchoidal, brittle
Crystal system: (Trigonal) short columnar
Chemical composition: CuSiO₂(OH)₂ hydrous copper silicate

Transparency: transparent, translucent
Refractive index: 1.644–1.709
Double refraction: +0.051 to +0.053
Dispersion: 0.036 (0.021)
Pleochroism: Weak; dark emerald green, light emerald green
Absorption spectrum: 550, 465
Fluorescence: None

So named (Greek—view through) because of its crystal structure. Deposits are found in Chile, Kyrgyzstan, Namibia, Peru, Russia, Arizona, and Zaire.
Possibilities for Confusion With demantoid (page 106), diopside (page 190), fluorite (page 198), emerald (page 90), uvarovite (page 106), and verdelite (page 110).

Apatite (group) [3–6]

Color: Colorless, pink, yellow, green, blue, violet
Color of streak: White to yellow-gray
Mohs' hardness: 5
Density: 3.16–3.23
Cleavage: Indistinct
Fracture: Conchoidal, brittle
Crystal system: (Hexagonal), columnar, thick tabular
Chemical composition: Ca₅(PO₄)₃(F,OH,Cl) basic fluoro- and chloro-calcium

phosphate
Transparency: Transparent
Refractive index: 1.628–1.649
Double refraction: −0.002 to −0.006
Dispersion: 0.013 (0.010)
Pleochroism: Green a.: yellow, green; Blue a.: very strong; blue, yellow
Absorption spectrum: Yellow and green a.: 597, 585, 577, 533, 529, 527, 525, 521, 514, 469; Blue a.: 512, 507, 491, 464
Fluorescence: Yellow a.: purple to pink

Name (Greek—cheat) because it can be easily confused. It has a vitreous luster; sensitive to acids. Deposits are found in Burma (Myanmar), Brazil, India, Kenya, Madagascar, Mexico, Norway, Sri Lanka, South Africa, and the United States. Apatite cat's-eye is known.
Possibilities for Confusion With amblygonite (page 192), andalusite (page 178), brazilianite (page 190), precious beryl (page 96), sphene (see below), topaz (page 102), and tourmaline (page 110). Synthetic apatite is known.
Asparagus stone Trade name for a light green apatite variety.

Sphene [7, 8] Also called Titanite

Color: Yellow, brown, green, reddish
Color of streak: White
Mohs' hardness: 5–5½
Density: 3.52–3.54
Cleavage: Good
Fracture: Conchoidal, brittle
Crystal system: (Monoclinic) platy
Chemical composition: CaTiSiO₅ calcium titanium silicate

Transparency: Transparent to opaque
Refractive index: 1.843–2.110
Double refraction: +0.100 to +0.192
Dispersion: 0.051 (0.019–0.038)
Pleochroism: Strong; colorless, greenish-yellow, reddish
Absorption spectrum: 586, 582
Fluorescence: None

The mineralogical name (titanite) derives from the titanium contents. It has an adamantine luster; with brilliant cut, it has an intensive fire. Deposits are found in Burma (Myanmar), Brazil, Mexico, Austria, Sri Lanka, and the United States. Titanite is changed to red or orange through heating.
Possibilities for Confusion With chrysoberyl (page 98), dravite (page 110), golden beryl (page 96), scheelite (page 196), topaz (page 102), zircon (page 108), and idocrase (page 186).

1 Dioptase, 2 crystals	5 Apatite, 8 different cuts
2 Dioptase, 12 faceted stones	6 Apatite, 3 crystals
3 Apatite, crystal	7 Sphene, 8 different cuts
4 Apatite cat's-eye, Brazil	8 Sphene, rough

194

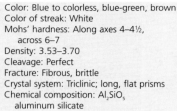

Kyanite [1–3] Also called Disthene

Color: Blue to colorless, blue-green, brown
Color of streak: White
Mohs' hardness: Along axes 4–4½,
    across 6–7
Density: 3.53–3.70
Cleavage: Perfect
Fracture: Fibrous, brittle
Crystal system: Triclinic; long, flat prisms
Chemical composition: Al$_2$SiO$_5$
    aluminum silicate

Transparency: Transparent, translucent
Refractive index: 1.710–1.734
Double refraction: –0.015
Dispersion: 0.020 (0.011)
Pleochroism: strong; colorless, blue,
    dark blue
Absorption spectrum: (706), (689), (671),
    (652), <u>446</u>, 433
Fluorescence: Weak; red

Name (Greek—blue) because of its color. It has a vitreous luster, often with irregular streaks. Difficult to cut because of variable hardness and cleavage. Deposits are found in Burma (Myanmar), Brazil, Kenya, Austria, Switzerland, Zimbabwe, and the United States.

Possibilities for Confusion With aquamarine (page 94), benitoite (page 184), iolite (page 180), dumortierite (page 182), sapphire (page 86), and tourmaline (page 110).

Scheelite [4, 5]

Color: Yellow, brown, orange, colorless
Color of streak: White
Mohs' hardness: 4½–5
Density: 5.9–6.3
Cleavage: Good
Fracture: Conchoidal, splintery, brittle
Crystal system: Tetragonal; dipyramids
Chemical composition: CaWO$_4$ calcium

    tungstate
Transparency: Transparent, translucent
Refractive index: 1.918–1.937
Double refraction: +0.010 to +0.018
Dispersion: 0.038 (0.026)
Pleochroism: Variable
Absorption spectrum: <u>584</u>
Fluorescence: Strong; light blue

Named after a Swedish chemist. It has an adamantine luster. Deposits are found in Japan, Korea, Mexico, Sri Lanka, and Arizona.

Possibilities for Confusion With chrysoberyl (page 98), diamond (page 70), golden beryl (page 96), and zircon (page 108). Synthetic scheelite has been known since 1963.

Variscite [6–8] Also called Utahlite

Color: Yellow-green, bluish
Color of streak: White
Mohs' hardness: 4–5
Density: 2.42–2.58
Cleavage: Perfect
Fracture: Conchoidal, brittle
Crystal system: Orthorhombic; short
    needles
Chemical composition: AlPO$_4$ • 2H$_2$O

    hydrous aluminum phosphate
Transparency: Translucent to opaque
Refractive index: 1.563–1.594
Double refraction: –0.031
Dispersion: Unknown
Pleochroism: Missing
Absorption spectrum: <u>688</u>, (650)
Fluorescence: Strong; pale green, green

So named (Latin—*variscia*) after old name for the Vogtland, Germany. Mostly exists as bulbous, coarse aggregates, frequently interspersed with brown matrix; is commonly cut en cabochon.

Possibilities for Confusion With chrysocolla (page 200), chrysoprase (page 128), jade (page 154), smaragdite (page 204), turquoise (page 170), and verdite (page 240).

Amatrix ("American matrix," also spelled amatrice and called variscite quartz)
Intergrowth of variscite with quartz or chalcedony, Nevada.

1 Kyanite, 3 broken crystals	5 Scheelite, 3 faceted stones
2 Kyanite, 5 different cuts	6 Variscite, 3 cabochons
3 Kyanite, crystal	7 Variscite in host rock, 2 pieces
4 Scheelite, broken crystal	8 Variscite with host rock, cabochon

Fluorite [1–3] Also called Fluorspar

Color: Colorless, all colors
Color of streak: White
Mohs' hardness: 4
Density: 3.00–3.25
Cleavage: Perfect
Fracture: Even to conchoidal, brittle
Crystal system: (Cubic) cubes, octahedra
Chemical composition: CaF_2 calcium fluoride

Transparency: Transparent, translucent
Refractive index: 1.434
Double refraction: None
Dispersion: 0.007 (0.004)
Pleochroism: Absent
Absorption spectrum: Green f.: 634, 610, 582, 445, 427
Fluorescence: Strong; blue-violet

Name (Latin—to flow) follows its being used as flux. It has a vitreous luster; color distribution is often zonal or spotty. Deposits are found in Oberpfalz, Bavaria (Germany), Argentina, Burma (Myanmar), England, France, Namibia, Austria, Switzerland, and Illinois.

Possibilities for Confusion   With many gemstones, due to its richness of color. Color can be changed with gamma rays. Synthetic fluorite is known in all colors.

Blue John   Fluorite variety, banded with colors and white, Derbyshire, England.

Hemimorphite [4–6] Also called Calamine

Color: Blue, green, colorless
Color of streak: White
Mohs' hardness: 5
Density: 3.30–3.50
Cleavage: Perfect
Fracture: Conchoidal, uneven, brittle
Crystal system: Orthorhombic; tabular
Chemical composition: $Zn_4Si_2O_7(OH)_2 \cdot H_2O$

hydrous basic zinc silicate
Transparency: Transparent to opaque
Refractive index: 1.614–1.636
Double refraction: +0.022
Dispersion: 0.020 (0.013)
Pleochroism: Absent
Absorption spectrum: Cannot be evaluated
Fluorescence: Weak; not characteristic

Named (Greek—half shape) because of its crystal formation. Aggregates are often blue-white banded, also mixed with dark matrix [no. 5]. Deposits are found in Algeria, Australia, Italy, Mexico, Namibia, Austria, and the United States.

Possibilities for Confusion   With chrysocolla (page 200), smithsonite (see below), and turquoise (page 170).

Smithsonite [7, 8] Also called Bonamite

Color: Light green, light blue, pink
Color of streak: White
Mohs' hardness: 5
Density: 4.00–4.65
Cleavage: Perfect
Fracture: Uneven, brittle
Crystal system: (Trigonal) rhombohedral
Chemical composition: $ZnCO_3$ zinc

carbonate
Transparency: Translucent, opaque
Refractive index: 1.621–1.849
Double refraction: –0.228
Dispersion: 0.014–0.031 (0.008–0.017)
Pleochroism: Absent
Absorption spectrum: Cannot be evaluated
Fluorescence: Blue-white, pink, brown

Named after an American mineralogist. It has a vitreous luster; grape-like clusters have a pearly luster, often slightly banded. Deposits are found in Australia, Greece, Italy (Sardinia), Mexico, Namibia, Spain, and New Mexico.

Possibilities for Confusion   With chrysoprase (page 128), hemimorphite (see above), jade (page 156), and turquoise (page 170).

1 Fluorite, 2 cleaved octahedrons	5 Hemimorphite, 3 cabochons
2 Fluorite, 2 rough pieces	6 Hemimorphite, radial aggregate
3 Fluorite, 9 different cuts	7 Smithsonite, 2 aggregates
4 Hemimorphite, crystal and 2 stones	8 Smithsonite, 3 cabochons

Sphalerite [1–3] Also called Zinc Blende

Color: Yellow, reddish, greenish, colorless	Transparency: Transparent, translucent
Color of streak: Yellowish to light brown	Refractive index: 2.368–2.371
Mohs' hardness: 3½–4	Double refraction: None
Density: 3.90–4.10	Dispersion: 0.156
Cleavage: Perfect	Pleochroism: Absent
Fracture: Uneven, brittle	Absorption spectrum: 690, 667, <u>651</u>
Crystal system: Cubic; tetrahedral	Fluorescence: Yellow-orange, red
Chemical composition: (Zn,Fe)S zinc sulphide	

Name (Greek—deceitful), due to the fact that it is used as an ore. It has a greasy and adamantine luster. Dispersion is three times as high as for diamond. Deposits are found in Canada, Mexico, Namibia, Spain, Wisconsin, and Zaire.

Possibilities for Confusion   With chrysoberyl (page 98), cassiterite (page 184), scheelite (page 196), sinhalite (page 186), topaz (page 102), tourmaline (page 110), idocrase (page 186), zircon (page 108), and colorless diamond (page 70). Syntheses are available in the trade.

Cerussite [4.5] Also called White-Lead Ore

Color: Colorless, gray, brownish	carbonate
Color of streak: White	Transparency: Transparent, translucent
Mohs' hardness: 3–3½	Refractive index: 1.804–2.079
Density: 6.46–6.57	Double refraction: –0.274
Cleavage: Good	Dispersion: 0.055 (0.033–0.050)
Fracture: Conchoidal, uneven, very brittle	Pleochroism: Absent
Crystal system: Orthorhombic; tabular, columnar	Absorption spectrum: Cannot be evaluated
	Fluorescence: Yellow, pink, green, bluish
Chemical composition: PbCO$_3$ lead	

Named (Latin—lead white) after its composition. It has an adamantine luster. It is very difficult to cut because of its high brittleness. Deposits are found in Australia, Italy, Austria, Madagascar, Namibia, Zambia, Scotland, and the United States.

Possibilities for Confusion   The same as for sphalerite (see above).

Chrysocolla [6–8]

Color: Green, blue	hydrous copper silicate
Color of streak: Green-white	Transparency: Opaque, sometimes just translucent
Mohs' hardness: 2–4	
Density: 2.00–2.40	Refractive index: 1.460–1.570
Cleavage: None	Double refraction: –0.023 to –0.040
Fracture: Conchoidal	Dispersion: None
Crystal system: Monoclinic; compact grape-like aggregates (botryoidal)	Pleochroism: Absent
	Absorption spectrum: Cannot be evaluated
Chemical composition: (Cu,Al)$_2$H$_2$Si$_2$O$_5$(OH)$_4$ • nH$_2$O	Fluorescence: None

It has a greasy vitreous luster. Deposits are found in Chile, Israel, Mexico, Peru, Russia, Nevada, and Zaire.

Possibilities for Confusion   With azurite (page 174), dyed chalcedony (page 126), malachite (page 176), turquoise (page 170), and variscite (page 196).

Chrysocolla quartz   Intergrowth of chrysocolla with quartz.

Eilat stone [8]   Intergrowth of chrysocolla with turquoise and malachite; found near Eilat/Israel.

1 Sphalerite, 3 rough pieces	5 Cerussite, 5 faceted stones
2 Sphalerite, faceted, 47.97ct	6 Chrysocolla, 4 cabochons
3 Sphalerite, 3 faceted stones	7 Chrysocolla, 2 rough pieces
4 Cerussite, twinned crystal	8 Eilat stone, 2 cabochons

Serpentine [1–3]

Color: Green, yellowish, brown	basic magnesium silicate
Color of streak: White	Transparency: semitransparent to opaque
Mohs' hardness: 2½–5½	Refractive index: 1.560–1.571
Density: 2.44–2.62	Double refraction: +0.008 to +0.014
Cleavage: None	Dispersion: None
Fracture: Uneven, splintery, tough	Pleochroism: Absent
Crystal system: (Monoclinic) microcrystalline	Absorption spectrum: Bowenite: 492, 464
Chemical composition: $H_4Mg_3Si_2O_9$	Fluorescence: Williamsite weak; greenish

There are two aggregate structures for serpentine (Latin—snake): leafy antigorite (leafy serpentine) and fibrous chrysotile (fibrous serpentine). Very finely fibrous varieties are called asbestos. It has a greasy to silky luster, and is sensitive to acids. Colors are often spotty. Deposits are found in Afghanistan, China, New Zealand, and the United States. Used mainly as decorative stone or for ornamental objects. There are many trade names for the "precious serpentine," which is sometimes misleadingly sold as jade.

Possibilities for Confusion With jade (page 154), onyx marble (page 218), turquoise (page 170), and verdite (page 240).

Bastite Silky shiny serpentine in the crystal shape of enstatite (page 192).

Bowenite Apple-green serpentine variety; frequently interspersed with light spots.

Connemara Green rock, intergrowth of serpentine with marble.

Verd-antique Green rock, interspersed with white calcite or dolomite veins; a serpentinite breccia.

Williamsite [3] Oil-green serpentine variety; often with black inclusions.

Stichtite [4]

It is pink-red to purple, formed through the decomposition of chrome-containing serpentinite. It has a vitreous luster. Deposits are found in Australia, Canada, Zimbabwe, and South Africa.

Ulexite [5, 6]

Color: White	$NaCaB_5O_6(OH)_6$ • $5H_2O$ hydrous sodium calcium borate
Color of streak: White	
Mohs' hardness: 2–2½	Transparency: Transparent, translucent
Density: 1.65–1.95	Refractive index: 1.491–1.520
Cleavage: Perfect	Double refraction: +0.029
Fracture: Fibrous	Dispersion: None
Crystal system: (Monoclinic) small; fibrous aggregates	Pleochroism: Absent
	Absorption spectrum: Cannot be evaluated
Chemical composition:	Fluorescence: Green-yellow, blue

Named after a German chemist. It has a silky luster. A piece of writing placed underneath the stone appears on the surface (therefore also called TV stone). Shows a cat's-eye effect [no. 5] when cut en cabochon. Deposits are found in Argentina, Chile, Canada, Kazakhstan, Peru, Russia, and the United States.

Tiger's-Eye Matrix [7, 8]

Trade name for a mineral aggregate, in which tiger's-eye–like structures (page 124) alternate with iron oxide layers.

1 Serpentine, rough	5 Ulexite, 3 cabochons
2 Chrysotile, 2 cabochons	6 Ulexite, 3 rough pieces
3 Williamsite, 2 faceted stones	7 Tiger's-eye matrix, 2 pieces partly polished
4 Stichtite, rough and cabochon	8 Tiger's-eye matrix, rough

Gemstones for Collectors

Many minerals can hardly be worn as jewelry because they are either too soft, too brittle, somehow or other endangered, or too rare. They are faceted, though, or offered en cabochon, for collectors and others who are interested in them. Compare also with the description on page 69.

1 **Gahnite** (Zinc spinel) Red-violet, also different blue shades, green or blackish. Transparent. Mohs' hardness 7½–8. Density 4.00–4.62. Cubic, $ZnAl_2O_4$. For other spinels, see page 100.

2 **Binghamite** Trade name for a quartz (page 116) with goethite inclusions. As cabochon, it shows a fine shimmer. Minnesota.

3 **Sanidine** [3 faceted stones] Light brownish orthoclase variety (page 164), also light gray. Transparent. Mohs' hardness 6. Strong, vitreous luster.

4 **Tantalite** [2 faceted stones] Red-brown, also brown or black. Transparent. Mohs' hardness 6–6½. Density 5.18–8.20. Orthorhombic, $(Fe,Mn)Ta_2O_6$.

5 **Rutile** [3 faceted stones] Reddish-brown, also blood-red or black. Usually opaque. Mohs' hardness 6–6½. Density 4.20–4.30. Tetragonal, TiO_2. Metallic adamantine luster.

6 **Peristerite** [4 cabochons] Opaque albite variety (plagiocase feldspar, page 164) with bluish shimmer on brown basic color.

7 **Haüynite** [3 faceted stones] Lazure blue, also green-blue, blue-white. Transparent to opaque. Mohs' hardness 5½–6. Density 2.4–2.5. Cubic, $(Na, Ca)_{8-4}Al_6Si_6(O,S)_{24}(SO_4,Cl)_{1-2}$. Vitreous luster.

8 **Tugtupite** (also called reindeer stone) [3 cabochons] Dark red with violet tint; it often has a spotty look because of mineral inclusions. Mostly opaque. Mohs' hardness 5½–6. Density 2.36–2.57. Tetragonal, $Na_4AlBeSi_4O_{12}Cl$. First discovered in 1960, South Greenland, Kola Peninsula (Russia).

9 **Willemite** [3 faceted stones] Yellow, green, red-brown, also white. Transparent to opaque. Mohs' hardness 5½. Density 3.89–4.18. Trigonal, Zn_2SiO_4. Crystals are rare. Resinous luster, strong green fluorescence.

10 **Natrolite** [3 faceted stones] Colorless, white, yellowish, also reddish and brownish. Transparent. Mohs' hardness 5–5½. Density 2.20–2.26. Orthorhombic, $Na_2(Al_2Si_3O_{10}) \bullet 2H_2O$.

11 **Smaragdite** [2 cabochons] Grass to emerald green. Translucent to transparent. Mohs' hardness 5½. Density 3.24–3.50. An actinolite variety (see below). It is mineralogically not related to "Smaragd" (emerald). An emerald imitation from fused masses is also called smaragdite.

12 **Leucite** [2 faceted stones] Colorless, white, yellowish. Transparent. Mohs' hardness 5½–6. Density 2.45–2.50. Tetragonal, $K(AlSi_2O_6)$.

13 **Actinolite** (also called ray stone) Green, also brown, gray, white, and colorless. Transparent. Mohs' hardness 5½–6. Density 3.03–3.07. Monoclinic, $Ca_2(Mg,Fe)_5Si_8O_{22}(OH)_2$.

14 **Hypersthene** [2 faceted stones] Black-green, black-brown. Transparent to opaque. Mohs' hardness 5–6. Density 3.4–3.5. Orthorhombic, $(Fe,Mg)_2Si_2O_6$.

15 **Datolite** [3 faceted stones] Greenish, yellow, also colorless. Transparent. Mohs' hardness 5–5½. Density 2.90–3.00. Monoclinic, $CaBSiO_4(OH)$.

1 **Periclase** [Cabochon and 2 faceted stones]   Yellowish, gray-green, also colorless. Transparent. Mohs' hardness 5½–6. Density 3.7–3.9. Cubic, MgO. Vitreous luster. Synthetic imitation for various gemstones.

2 **Purpurite**   Purple-colored, deep pink, also dark brown. Translucent. Mohs' hardness 4–4½. Density 3.2–3.4. Orthorhombic, $MnPO_4$. Metallic luster, brittle.

3 **Apophyllite**   Colorless or weakly red, also yellowish, greenish to bluish. Transparent. Mohs' hardness 4½–5. Density 2.30–2.50. Tetragonal, $KFCa_4(Si_2O_5)_4 \bullet 8H_2O$. Mother-of-pearl luster and characteristic light sheen. (See also picture on page 189, no. 3.)

4 **Zincite** (also called red-zinc ore) [3 faceted stones]   Red to orange-red. Translucent. Mohs' hardness 4–5. Density 5.66. Hexagonal, (Zn, Mn)O. Greasy to adamantine luster, brittle. Zincite syntheses are on the market.

5 **Kurnakovite** [2 faceted stones]   Colorless, also pink. Transparent. Mohs' hardness 3. Density 1.86. Triclinic, $MgB_3O_3(OH)_5 \bullet 5H_2O$.

6 **Siderite** (also called chalybite and ironspar) [2 faceted stones]   Red-brown and gold-brown. Transparent. Mohs' hardness 3½–4½. Density 3.83–3.96. Trigonal, $FeCO_3$. Radial, globular aggregates are cut en cabochon.

7 **Colemanite** [3 faceted stones]   Colorless, gray-white. Transparent. Mohs' hardness 4½. Density 2.40–2.42. Monoclinic, $Ca_2B_6O_{11} \bullet 5H_2O$. Strong, vitreous luster.

8 **Cuprite** (also called red-copper ore)   Carmine-red. Translucent. Mohs' hardness 3½–4. Density 5.85–6.15. Cubic, Cu_2O. Metallic luster; very high refractive index. There are various types of stone offered in the trade as cuprite which contain very little cuprite but are merely dyed red.

9 **Barite** (also called barytes and heavy spar) [5 faceted stones]   Colorless, brown, yellow, also red, green, blue. Transparent to opaque. Mohs' hardness 3–3½. Density 4.43–4.46. Orthorhombic, $BaSO_4$. Vitreous luster, brittle.

10 **Dolomite** [3 faceted stones]   Colorless, also pastel-colored. Transparent. Mohs' hardness 3½–4. Density 2.80–2.95. Trigonal, $CaMg(CO_3)_2$. Vitreous luster.

11 **Chalcopyrite** [also called copper pyrite]   Brass-yellow, gold-yellow, with a green tint. Opaque: Mohs' hardness 3½–4. Density 4.10–4.30. Tetragonal, $CuFeS_2$. Metallic luster.

12 **Witherite** [2 faceted stones]   Yellowish-white to colorless. Transparent. Mohs' hardness 3–3½. Density 4.27–4.35. Orthorhombic, $BaCO_3$. Waxy, vitreous luster. Witherite dust is toxic; do not inhale.

13 **Anhydrite**   Colorless, bluish, also red-violet. Transparent. Mohs' hardness 3½. Density 2.90–2.98. Orthorhombic, $CaSO_4$. Bright luster.

14 **Magnetite-jade**   Trade name for opaque, black jade with galvanic gilded, originally also black, magnetite inclusions. Mohs' hardness 5½–7. Density 3.4–4.4. Deposits in California. (For jade, see page 154.)

1 **Calcite** (also called limespar) [3 faceted stones]   Colorless, yellow, brown, and various other colors. Transparent to translucent. Mohs' hardness 3. Density 2.69–2.71. Trigonal, $CaCO_3$. Vitreous luster.

2 **Howlite**   Snow-white, occasionally dispersed with black or dark-brown veins. Mohs' hardness 3–3½. Density 2.45–2.58. Monoclinic, $Ca_2B_5SiO_9(OH)_5$. Easily dyed as it is very porous (see below, no. 16).

3 **Cobalti-calcite** [Cabochon and faceted]   Calcite, colored violet-red with cobalt (see above, no. 1). Cuttable material from Spain.

4 **Barytocalcite**   Yellowish-white.   Transparent to translucent. Mohs' hardness 4. Density 3.66. Monoclinic, $BaCa(CO_3)_2$. Vitreous luster, brittle.

5 **Celestite** (also called celestine) [3 faceted stones]   Bluish-white, colorless, seldom also reddish, green. Transparent. Mohs' hardness 3–3½. Density 3.97–4.00. Orthorhombic, $SrSO_4$. Vitreous luster, brittle.

6 **Wulfenite** (also called yellow-lead ore)   Honey-yellow, orange, also red. Transparent to translucent. Mohs' hardness 3. Density 6.50–7.00. Tetragonal, $PbMoO_4$. Resinous to adamantine luster, brittle. Very sensitive to heat.

7 **Aragonite** [2 faceted stones]   Colorless, green, also other colors. Transparent. Mohs' hardness 3½–4. Density 2.94. Orthorhombic, $CaCO_3$. Vitreous luster. Onyx marble (page 218), ammonite (page 240), pearls (page 230), and mother-of-pearl (page 239) consist totally, or to a large extent, of aragonite.

8 **Crocoite** (also called red-lead ore) [3 faceted stones]   Red-orange, also yellow, transparent to translucent. Mohs' hardness 2½–3. Density 5.9–6.1. Monoclinic, $PbCrO_4$. Adamantine luster.

9 **Gaylussite**   White, also colorless, yellow. Transparent. Mohs' hardness 2½–3. Density 1.99. Monoclinic, $Na_2Ca(CO_3)_2 \bullet 5H_2O$. Dull, vitreous luster.

10 **Phosgenite** (also called lead-horn ore) [4 faceted stones]   Colorless, white, yellowish, also greenish. Transparent. Mohs' hardness 2–3. Density 6.13. Tetragonal, $Pb_2(CO_3)Cl_2$. Greasy, adamantine luster.

11 **Silver**   The originally silver-white dendritic inclusions in quartz that have turned deep black. Opaque. Mohs' hardness 2½–3. Density 9.6–12.0. Cubic, Ag. Metallic luster.

12 **Gold**   Pure gold included in quartz. Golden-yellow. Opaque. Mohs' hardness 2½–3. Density 15.5–19.3. Cubic, Au. Metallic luster.

13 **Vivianite**   Blue-green, also deep blue, colorless. Transparent to translucent. Mohs' hardness 1½–2. Density 2.64–2.70. Monoclinic, $Fe_3(PO_4)_2 \bullet 8H_2O$. Vitreous to mother-of-pearl luster. Very strong pleochroism.

14 **Sulfur**   Yellow, also brownish. Translucent. Mohs' hardness 1½–2½. Density 2.05–2.08. Orthorhombic, S. adamantine luster. Extremely sensitive to heat, bursts when warmed in one's hand.

15 **Proustite**   Cinnabar to scarlet-red. Translucent. Mohs' hardness 2½. Density 5.51–5.64. Trigonal, Ag_3AsS_3. Adamantine luster. The color gradually darkens when exposed to light.

16 **Dyed Howlite**   When colored blue, howlite (see above, no. 2) is occasionally used to imitate turquoise (page 170).

1 **Boleite**   Indigo-blue, also deep blue. Transparent to translucent. Mohs' hardness 3–3½. Density 5.05. Cubic. $Pb_{26}Ag_{10}Cu_{24}Cl_{62}(OH)_{48} \cdot 3H_2O$. Vitreous luster; on cleavage surfaces mother-of-pearl luster

2 **Oligoclase**   [3 faceted stones]   Colorless to bluish plagioclase feldspar. Transparent. Mohs' hardness 6–6½. Density 2.62–2.67. Triclinic, $(Na,Ca)(Si,Al)_2Si_2O_8$. Vitreous luster, on cleavage surfaces mother-of-pearl luster.

3 **Ludlamite**   Light- to apple-green, colorless. Transparent to translucent. Mohs' hardness 3–4. Density 3.1–3.2. Monoclinic, $(Fe,Mg,Mn)_3(PO_4)_2 \cdot 4H_2O$. Vitreous luster.

4 **Adamite** [2 faceted stones]   Yellow-green, brownish, also colorless, pink, violet. Transparent to translucent. Mohs' hardness 3½. Density 4.30–4.68. Orthorhombic, $Zn_2(AsO_4)(OH)$. Strong, vitreous luster.

5 **Augelite**   Colorless to slightly yellowish. Transparent. Mohs' hardness 4½–5. Density 2.70–2.75. Monoclinic, $Al_2(PO_4)(OH)_3$.

6 **Friedelite**   Pink-red to red-brown. Translucent to opaque. Mohs' hardness 4–5. Density 3.06–3.19. Monoclinic (pseudo-trigonal), $Mn_8Si_6O_{15}(OH,Cl)_{10}$. Vitreous luster.

7 **Talc** [2 cabochons]   Gray-green, also pearl-white, blue-green, yellowish. Translucent to opaque. Mohs' hardness 1. Density 2.55–2.80. Monoclinic (pseudohexagonal), $Mg_3Si_4O_{10}(OH)_2$. Mother-of-pearl-luster, greasy luster; also dull. Feels greasy. Dense aggregates are called soapstone or steatite; they can be easily carved and are worked on the wheel into ornamental objects.

8 **Manganotantalite** [3 faceted stones]   Scarlet-red to dark red. Transparent. Mohs' hardness 5½–6½. Density 7.73–7.97. Orthorhombic, $MnTa_2O_6$. Vitreous luster.

9 **Gadolinite**   Black, also green-black, brown, very rarely pale green. Usually opaque; light varieties are transparent. Mohs' hardness 6½–7. Density 4.00–4.65. Monoclinic, $Y_2FeBe_2Si_2O_{10}$. Vitreous to greasy luster; very brittle.

10 **Anglesite** [3 faceted stones]   Colorless, yellow, also greenish, brownish, white. Transparent to translucent. Mohs' hardness 3–3½. Density 6.30–6.39. Orthorhombic, $PbSO_4$. Vitreous to adamantine luster; brittle.

11 **Vlasovite** [3 faceted stones] Yellow, yellow-brown, also colorless. Transparent. Mohs' hardness 6. Density 2.97. Monoclinic, $Na_4Zr_2Si_4O_{11}$. Greasy luster; on cleavage surfaces vitreous to mother-of-pearl-luster; brittle.

12 **Ekanite**   Dark green, also yellow, light to dark brown. Transparent to translucent. Mohs' hardness 6–6½. Density 3.28–3.32. Originally tetragonal, later amorphous, due to persisting radioactivity, $ThCa_2Si_8O_{20}$. The data of crystalline ekanite is somewhat different from that of the amorphous variety. Vitreous luster. *Radioactive!*

13 **Phosphophyllite** [2 faceted stones]   Blue-green, also colorless. Transparent. Mohs' hardness 3–3½. Density 3.7–3.13. Monoclinic, $Zn_2(Fe,Mn)[PO_4]_2 \cdot 4H_2O$. Vitreous luster. Perfect cleavage; brittle.

14 **Gypsum** (also called selenite) [2 cabochons]   White, yellowish, also colorless, pink, bluish. Transparent to opaque. Mohs' hardness 2. Density 2.20–2.40. Monoclinic, $CaSO_4 \cdot 2H_2O$. Vitreous luster; fibrous aggregates show silky luster. Very sensitive to heat. Gypsum is rock-forming; compare to alabaster, page 222.

The illustrations are 40 percent larger than the originals.

1 **Analcite Cat's-Eye**   Pink, South Africa. (For more data, see below, no. 2.)

2 **Analcite** [analcime] [4 faceted stones]   Colorless, white, also pink, yellowish, greenish. Transparent to opaque. Mohs' hardness 5–5½. Density 2.22–2.29.

3 **Triphylite** [3 faceted stones]   Brownish, also greenish-gray, bluish-gray. Transparent to translucent. Mohs' hardness 4–5. Density 3.34–3.58. Orthorhombic, $LiFePO_4$. Vitreous luster; brittle.

4 **Staurolite**   Reddish-brown, black. Transparent to opaque. Mohs' hardness 7–7½. Density 3.7–3.8. Monoclinic. Vitreous Luster. Characteristic interpenetration twinning.

5 **Hornblende** [2 faceted stones]   Brown-yellow, also green, black. Translucent to opaque. Mohs' hardness 5–6. Density 2.9–3.4. Monoclinic.

6 **Pectolite** [3 faceted stones]   Light blue to light green, also colorless, gray. Transparent to translucent. Mohs' hardness 4½–5. Density 2.74–2.88. Triclinic, $NaCa_2Si_3O_8(OH)$. Vitreous to silky luster.

7 **Zektzerite** [2 faceted stones]   Colorless, also light pink. Transparent to translucent. Mohs' hardness 6. Density 2.79. Orthorhombic.

8 **Nepheline** [3 faceted stones]   Light pink, also colorless, white, greenish, yellowish. Transparent to opaque. Mohs' hardness 5–6. Density 2.55–2.65. Hexagonal, $(Na,K)AlSiO_4$. Vitreous luster; on surface of fracture greasy luster. Elaeolite is a dull, mostly greenish nepheline variety.

9 **Greenockite** [synthetic]   Orange, also yellow to brown. Transparent to translucent. Mohs' hardness 3–3½. Density 4.73–4.79. Hexagonal, CdS.

10 **Anatase** [2 faceted stones]   Dark brown, also colorless, yellow, blue, reddish, black. Transparent to translucent. Mohs' hardness 5½–6. Density 3.82–3.97. Tetragonal, TiO_2. Adamantine luster.

11 **Milarite**   Yellow, also colorless, white, green. Transparent. Mohs' hardness 5½–6. Density 2.46–2.61. Hexagonal, $KCa_2AlBe_2Si_{12}O_{30} \bullet 0.5H_2O$. Vitreous luster.

12 **Descloizite**   Red-brown, also orange, brown-black. Transparent to translucent. Mohs' hardness 3–3½. Density 5.5–6.2. Orthorhombic.

13 **Lithiophilite** [2 faceted stones]   Brown, also yellow, blue. Transparent to translucent. Mohs' hardness 4–5. Density 3.4–3.6. Orthorhombic.

14 **Whewellite**   Yellowish, faceted. (For more data, see below, no. 20.)

15 **Jeremejevite** [2 faceted stones]   Bluish, also colorless, yellowish. Transparent. Mohs' hardness 6½–7½. Density 3.28–3.31. Hexagonal, $Al_6B_5O_{15}(F,OH)_3$.

16 **Clinohumite** [2 faceted stones]   Orange to yellow, also white. Transparent to opaque. Mohs' hardness 6. Density 3.13–3.75. Monoclinic.

17 **Clinozoisite** [3 faceted stones]   Yellow to yellow-brown, also colorless, greenish, pink. Transparent to translucent. Mohs' hardness 6–7. Density 3.21–3.38. Monoclinic, $Ca_2Al_3(SiO_4)_3(OH)$. Vitreous luster.

18 **Eudialyte** [2 cabochons]   Brown-red, brown, also pink. Translucent to opaque. Mohs' hardness 5–5½. Density 2.74–2.98. Trigonal.

19 **Neptunite** [2 faceted stones]   Black. Translucent to opaque. Mohs' hardness 5–6. Density 3.19–3.23. Monoclinic. Strong, vitreous luster.

20 **Whewellite** [3 faceted stones]   Colorless, white, yellowish, also brownish. Transparent. Mohs' hardness 2½–3. Density 2.19–2.25. Monoclinic, $CaC_2O_4 \bullet H_2O$. Vitreous to mother-of-pearl-luster. Whewellite is an organic product.

The illustrations are 20 percent larger than the originals.

1 **Montebrasite** [5 faceted stones]   Colorless, yellowish, light green, light blue, also white, pink. Transparent. Mohs' hardness 5½–6. Density 2.98–3.11. Triclinic, $LiAl(PO_4)(OH,F)$. Vitreous luster.

2 **Cinnabar** [2 faceted stones]   Red, also pale blue. Translucent to opaque. Mohs' hardness 2–2½. Density 8.0–8.2. Trigonal, HgS. Adamantine luster.

3 **Boracite**   Light green, also colorless, white, yellow, bluish. Transparent to translucent. Mohs' hardness 7–7½. Density 2.95–2.96. Orthorhombic, $Mg_3B_7O_{13}Cl$. Vitreous to adamantine luster.

4 **Magnesite**   Colorless, also white, yellow to brown. Transparent to translucent. Mohs' hardness 3½–4½. Density 2.96–3.12. Trigonal, $MgCO_3$. Vitreous to dull luster.

5 **Wolframite**   Black, also dark brown. Translucent to opaque. Mohs' hardness 5–5½. Density 7.1–7.6. Monoclinic, $(Fe,Mn)WO_4$. Metallic to greasy luster. Wolframite is a mixed crystal of ferberite and hübnerite.

6 **Herderite** [2 faceted stones]   Grayish blue, also colorless, pale green, yellowish. Transparent to translucent. Mohs' hardness 5–5½. Density 2.95–3.02. Monoclinic, $CaBe(PO_4)F$. Vitreous luster.

7 **Leucophane**   Yellow with needle-like aegirine inclusions, also greenish. Transparent. Mohs' hardness 4. Density 3.0. Triclinic $(Na,Ca)_2BeSi_2O_6(F,OH)$.

8 **Pyrargyrite**   Dark red. Translucent. Mohs' hardness 2½–3. Density 5.85. Trigonal, Ag_3SbS_3. Adamantine luster.

9 **Bustamite**   Light pink, also brown-red. Transparent. Mohs' hardness 5½–6. Density 3.32–3.43. Triclinic, $(Mn,Ca)_3Si_3O_9$. Vitreous luster.

10 **Magnetite** [also called lodestone]   Black, opaque. Mohs' hardness 5½–6½. Density 5.2. Cubic, Fe_3O_4. Metallic luster; strongly magnetic. The originally bright black color gradually changes to dark brown.

11 **Strontianite**   Light yellow, also colorless, white, brown, green, reddish. Transparent to translucent. Mohs' hardness 3½. Density 3.63–3.79. Orthorhombic, $SrCO_3$. Vitreous luster; on surface of fracture, greasy luster.

12 **Parisite** [3 faceted stones]   Yellow-brown, also reddish. Transparent to translucent. Mohs' hardness 4½. Density 4.33–4.42. Trigonal, $Ca(Ce,La)_2(CO_3)_3F_2$. Vitreous luster.

13 **Eosphorite**   Yellow-brown, also colorless, pink. Transparent to translucent. Mohs' hardness 5. Density 3.05–3.08. Monoclinic, $MnAl(PO_4)(OH)_2•H_2O$. Vitreous luster.

14 **Senarmontite**   Colorless, also white, gray. Transparent to translucent. Mohs' hardness 2–2½. Density 5.2–5.5. Cubic, Sb_2O_3. Adamantine luster.

15 **Taaffeite**   Violet, also colorless, pale green, bluish, pink, red. Transparent. Mohs' hardness 8–8½. Density 3.60–3.62. Hexagonal, $Mg_3Al_8BeO_{16}$. Vitreous luster.

16 **Simpsonite**   Orange, also colorless, white, brownish-yellow. Translucent. Mohs' hardness 7–7½. Density 5.92–6.84. Trigonal, $Al_4(Ta,Nb)_3(O,OH,F)_{14}$. Vitreous to adamantine luster.

17 **Diaspore**   Greenish-brown, also colorless, white, yellow, bluish, pink. Transparent to translucent. Mohs' hardness 6½–7. Density 3.30–3.39. Orthorhombic, $AlO(OH)$. Vitreous luster; on cleavage surfaces mother-of-pearl luster.

The illustrations are 40 percent larger than the originals.

1 **Thaumasite** [3 faceted stones] Colorless, light yellow, also white. Translucent. Mohs' hardness 3½. Density 1.91. Hexagonal, $Ca_6Si_2(CO_3)_2(SO_4)_2(OH)_{12} \cdot 24H_2O$. Vitreous luster.

2 **Cancrinite** [4 faceted stones, 1 cabochon] Yellow, orange, also colorless, white, bluish, pink. Transparent to translucent. Mohs' hardness 5–6. Density 2.42–2.51. Hexagonal, $Na_2Ca_2Al_6Si_6O_{24}(CO_3)_2$. Vitreous luster; on cleavage surfaces mother-of-pearl luster.

3 **Tremolite** (also called grammatite) [6 faceted stones, 1 cabochon] Gray-brown, green, also colorless, white, pink. Transparent to translucent. Mohs' hardness 5–6. Density 2.95–3.07. Monoclinic, $Ca_2Mg_5Si_8O_{22}(OH)_2$. Vitreous to silky luster.

4 **Yugawaralite** [2 faceted stones] Colorless, also dull white. Transparent to translucent. Mohs' hardness 4½. Density 2.19–2.23. Monoclinic, $Ca(Al_2Si_6O_{16}) \cdot 4H_2O$. Vitreous luster.

5 **Sapphirine** [2 faceted stones] Dark blue, also colorless, greenish, pink, violet. Transparent. Mohs' hardness 7½. Density 3.40–3.58. Monoclinic, $(Mg,Al)_8(Al,Si)_6O_{20}$. Vitreous luster. So named because of its blue color similar to sapphire.

6 **Aegirine-augite** Black, also green-black. Transparent. Mohs' hardness 6. Density 3.40–3.55. Monoclinic. Vitreous luster. Aegirin is an augite that is rich in sodium.

7 **Melinophane** (also called meliphanite) [2 faceted stones] Honey-yellow, orange, also colorless, yellowish-red. Translucent. Mohs' hardness 5–5½. Density 3.00–3.03. Tetragonal $(Ca,Na)_2Be(Si,Al)_2(O,OH,F)_7$. Vitreous luster.

8 **Pollucite** (also called pollux) [3 faceted stones] Colorless, gray, also white, bluish, violet. Transparent to translucent. Mohs' hardness 6½–7. Density 2.85–2.94. Cubic $(Cs,Na)(AlSi_2O_6) \cdot ½H_2O$. Vitreous luster.

9 **Andesine** Light pink, also white, gray, yellowish, greenish. Transparent to translucent. Mohs' hardness 6–6½. Density 2.65–2.69. Triclinic $(Na,Ca)(Si,Al)_2Si_2O_8$. Vitreous to dull luster; on cleavage surfaces mother-of-pearl luster. Perfect cleavage. Andesine belongs to the group of plagioclase feldspars (see page 164).

10 **Legrandite** Yellow, also colorless. Transparent. Mohs' hardness 4½–5. Density 3.98–4.04. Monoclinic, $Zn_2(AsO_4)(OH) \cdot H_2O$. Vitreous luster.

11 **Muscovite** Pink, also colorless, silver-white, yellowish, greenish. Transparent to translucent. Mohs' hardness 2–3. Density 2.78–2.88. Monoclinic, $KAl_2AlSi_3O_{10}(OH)_2$. Vitreous luster; on cleavage surfaces mother-of-pearl luster. Perfect cleavage. Because of its silvery sheen, muscovite is called cat's silver in the vernacular. A chrome-containing muscovite variety, which is green, is called fuchsite.

12 **Davidite** Black. Transparent. Mohs' hardness 5–6. Density 4.5. Trigonal $(Ce,La)(Y,U,Fe)(Ti,Fe)_{20}(O,OH)_{38}$. Metallic luster.

13 **Mesolite** Colorless, also white. Transparent to translucent. Mohs' hardness 5–5½. Density 2.26–2.40. Orthorhombic, $Na_{16}Ca_{16}[Al_{48}Si_{72}O_{240}] \cdot 64H_2O$. Vitreous to dull luster.

14 **Pyrolusite** Black, also dark gray. Transparent. Mohs' hardness 6–7. Density 4.5–5.0. Tetragonal, MnO_2. Strong metallic luster. Color can rub off on fingers.

The illustrations are 50 percent larger than the originals.

Rocks as Gemstones

Formerly, rocks were used almost exclusively for decorative purposes and for ornamental objects. Nowadays, rocks are becoming more and more important also for personal jewelry, especially for costume jewelry. (Refer also to page 69.)

Onyx Marble [1, 2] Also called Marble Onyx

Color: Yellow-green, white, brown, striped	Composition: Calcite or aragonite
Color of streak: According to color of stone	Transparency: Translucent to opaque
Mohs' hardness: 3½–4	Refractive index: 1.486–1.686
Density: 2.72–2.85	Double refraction: –0.156 to –0.172

The rock that is offered in the trade as onyx marble is a limestone formed by the minerals calcite or aragonite (page 208). It must not be confused with chalcedony-onyx (page 142). It is incorrect to name this rock onyx without the addition of the word *marble*. Onyx marble is formed from lime-containing water by layered deposits (therefore always banded) near warm springs or as stalactites or stalagmites in caves. Deposits are found in Egypt, Algeria, Argentina, Morocco, Mexico, and the United States. It is used for ornamental objects such as pendants and brooches. Dyeing produces a wide variety of colors.

Possibilities for Confusion With the mineral serpentine (page 202) as well as with various serpentine rocks, especially with the green-white spotted Connemara from Connaught, Ireland (compare with page 202) or with the banded, greenish ricolite from New Mexico, United States.

Mexican onyx Misleading trade name for onyx marble.

Tufa [5–7] Also called Aragonite Sinter

Color: White, yellow, brown, reddish	Composition: Aragonite
Color of streak: According to color of stone	Transparency: Translucent, opaque
Mohs' hardness: 3½–4	Refractive index: Aragonite 1.530–1.685
Density: 2.95	Double refraction: Aragonite –0.155

Tufa is the calcium carbonate deposit from hot springs in the form of encrustations or stalactites, often with wavy layers. The best-known occurrence is at Karlsbad (Karlovy Vary) in the Czech Republic. Other deposits are found in Argentina, Mexico, New Zealand, Russia, and the United States.

Landscape Marble [8] Also called Ruin Marble
This is a fine-grained limestone, where the layers have been fractured, displaced, and again solidified. Because of the different coloring of the individual layers, images are formed which convey the impression of a landscape. It is used as a decorative stone, also for brooches or pendants, cut en cabochon. The landscape marble shown on the right could make one think of skyscrapers in a metropolis with low, thundery clouds.

1 Onyx marble, bowl
2 Onyx marble, broken piece, partly polished
3 Onyx marble, 2 pendants
4 Onyx marble, figurine
5 Tufa, 2 pieces from Karlsbad, Czech Rep.
6 Tufa brooch and pendant
7 Tufa, New Mexico, United States
8 Landscape marble, Tuscany, Italy
The illustrations are 50 percent smaller than the originals.

Orbicular Diorite [1, 2] Also called Ball Diorite

Plutonic rock composed of feldspar, hornblende, biotite, and quartz. Formed by rythmic crystallization which produces a separation of light and dark materials into spherical shapes. Used as a decorative stone, also cut en cabochon [no. 2].

Kakortokite

Trade name for a black-white-red banded nepheline-syenite from Greenland, containing eudialyte.

Obsidian [3-7]

Color: Black, gray, brown, green	Transparency: Transparent to opaque
Color of streak: White	Refractive index: 1.45–1.55
Mohs' hardness: 5–5½	Double refraction: None
Density: 2.35–2.60	Dispersion: 0.010
Cleavage: None	Pleochroism: Absent
Fracture: Large conchoidal, sharp edged	Absorption spectrum: Not diagnostic
Composition: Volcanic, amorphous, siliceous glassy rock	Fluorescence: None

Obsidian (named after the Roman, Obsius) was used in antiquity for amulets and necklaces. Varieties show golden [Gold o., no. 4] or silver [Silver o., no. 5] sheen, caused by inclusions. Deposits are found in Ecuador, Indonesia, Iceland, Italy, Japan, Mexico, and the United States.

Possibilities for Confusion With aegirine-augite (page 216), gadolinite (page 210), jet (page 226), hematite (page 162), pyrolusite (page 216), and wolframite (page 214).

Snowflake obsidian [6, 7] Trade name for an obsidian with gray-white, ball-shaped inclusions (spherulites). Deposits are found in Mexico and the United States.

Moldavite [10-12] Also called Bouteille Stone

Color: Bottle-green to brown-green	dioxide + (aluminum oxide)
Color of streak: White	Transparency: Transparent to opaque
Mohs' hardness: 5½	Refractive index: 1.48-1.54
Density: 2.32–2.38	Double refraction: None
Cleavage: None	Dispersion: None
Fracture: Conchoidal	Pleochroism: Absent
Crystal system: Amorphous	Absorption spectrum: Not diagnostic
Chemical composition: $SiO_2(+Al_2O_3)$ silicon	Fluorescence: None

Moldavite (named after Moldau [Vltava], Czech Republic) belongs to the tektite group; possibly formed from condensed rock vapors after being hit by a meteorite. Scarred surfaces; vitreous luster.

Possibilities for Confusion With apatite (page 194), diopside (page 190), precious beryl (page 96), sapphire (page 86), tourmaline (page 110), and with green bottle glass. Other tektites are dark brown to black [8, 9]. Depending on the place of discovery, they have different names, e.g., Australite (Australia), Billitonite (Borneo), Georgiaite (Georgia, United States), Indochinite (Indochina), Javaite (Java), and Philippinite (the Philippines).

1 Orbicular diorite, partly polished, Corsica	7 Snowflake obsidian, partly polished
	8 Tektite, rough, Thailand
2 Orbicular diorite, cabochon, Corsica	9 Tektite, 2 faceted stones, Thailand
3 Obsidian, rough, Mexico	10 Moldavite, 4 rough pieces, Czech Republic
4 Golden obsidian, Mexico	11 Moldavite, 6 faceted stones,
5 Silver obsidian, Mexico	Czech Republic
6 Snowflake obsidian, 2 cabochons	12 Moldavite, cabochon, Czech Republic
The illustrations are 40 percent smaller than the originals.	

Alabaster [1, 2]

Color: White, pink, brownish
Color of streak: White
Mohs' hardness: 2
Density: 2.30–2.33
Chemical composition: $CaSO_4 \cdot 2H_2O$

hydrous calcium sulfate
Transparency: Opaque, translucent at edges
Refractive index: 1.520–1.530
Double refraction: +0.010
Dispersion: None

Alabaster (from Greek) is the fine-grained variety of gypsum (page 210); in antiquity, the name also referred to microcrystalline limestone. Deposits are found in Germany (Thuringia), England (Derbyshire), France (Parisian Basin), Italy (Tuscany), and the United States (Colorado). Used for ornamental objects, rarely as jewelry. Can easily be dyed because of its porosity.

Possibilities for Confusion   With agalmatolite (see below), calcite (page 208), gypsum (page 210), meerschaum (see below), and onyx marble (page 218).

Agalmatolite Also called Picture Stone, Pagoda Stone, Pagodite

Whitish, greenish, or yellowish, dense aggregate of the mineral pyrophyllite; $Al_2Si_4O_{10}(OH)_2$. When heat-treated, the originally soft stone (Mohs' hardness 1–1½) hardens considerably. Deposits are found in Finland, Slovakia, South Africa, and the United States (California). Usage as for alabaster (see above).

Possibilities for Confusion   With alabaster (see above), talc (page 210). The greenish variety is used to imitate jade (page 154).

Meerschaum [3-5] Also called Sepiolite

Color: White, also yellowish, gray, reddish
Color of streak: White
Mohs' hardness: 2–2½
Density: 2.0–2.1
Cleavage: Perfect
Fracture: Flat conchoidal, earthy
Crystal system: Orthorhombic;
    microcrystalline
Chemical composition: $Mg_4Si_6O_{15}(OH)_2$

$\cdot 6H_2O$ hydrous magnesium silicate
Transparency: Opaque
Refractive index: 1.53
Double refraction: None
Dispersion: None
Pleochroism: Absent
Absorption spectrum: Cannot be evaluated
Fluorescence: None

Because of its high porosity, meerschaum can float (German—sea foam). It occurs as a rock as concretion in serpentine. It has a dull greasy luster, feels like soap, and sticks to the tongue. The most important deposit is near Eskişşehir, Anatolia (Turkey). It is also found in Greece (Samos), Morocco, Spain, Tanzania, and the United States (Texas). Worked into bowls for pipes and cigarette holders, which, because of the smoke, gradually become golden-yellow [no. 5]; in addition, it is used for costume jewelry. Becomes lustrous when impregnated with grease. Can be confused with alabaster (see above).

Fossils [6-9]

Petrified wood pieces (page 148) and other fossils (petrified animals or parts of animal) are attractive for making jewelry because of their form, structure, color, as well as their old age. For more about their formation, see page 148.

---

1 Alabaster, 2 pieces dyed red	6 Ammonite, shell replaced by pyrite
2 Alabaster ashtray, dyed blue	7 Ammonite, shell replaced by pyrite, partly polished
3 Meerschaum, rough	8 Trilobite, primeval crab in shale
4 Meerschaum, costume jewelry	9 Actaeonella, a sea snail, Austria, partly polished
5 Meerschaum, cigarette holder	

The illustrations are 20 percent smaller than the originals.

Organic Gemstones

A number of gemstones are of organic origin, but have the quality and character of stones. Refer also to the description on page 69.

Coral

Color: Red, pink, white, black, blue	Transparency: Translucent, opaque
Color of streak: White	Refractive index: White and red:
Mohs' hardness: 3–4	1.486–1.658
Density: White and red: 2.60–2.70	Double refraction: White and red: –0.160
Cleavage: None	to –0.172
Fracture: Irregular, splintery, brittle	Dispersion: None
Crystal system: (Trigonal) microcrystalline	Pleochroism: Absent
Chemical composition: $CaCO_3$ or organic	Absorption spectrum: Not diagnostic
substance	Fluorescence: Weak; violet

The coral (name of Greek origin) used as gem material is a branching skeleton-like structure built by small marine animals (coral polyps) related to reef-forming corals. The height of the branches is in the general range 8–10 in (20–40 cm); the branches are up to 2½ in (6 cm) thick.

Deposits are found along the coasts of the western Mediterranean countries, the Red Sea, Bay of Biscay, Canary Islands, Malaysian Archipelago, the Midway Islands, Japan, and Hawaii (United States). Production is increasingly controlled by environmental laws.

The coral is found at depths of 1–1020 ft (3–300 m), and mainly harvested with weighted, wide-meshed nets dredged across the seabed. When harvested by divers, however, not as many corals are damaged. Near Hawaii, minisubmarines have recently been used to collect the coral. When it is brought to the surface, the soft parts are rubbed away and the material is sorted as to quality. For more than 200 years, the main trade center has been Torre del Greco, south of Naples, Italy. Three-quarters of the corals harvested all over the world are still processed here.

Unworked coral is dull; when polished it has a vitreous luster. It is polished with fine-grained sandstone and emery; finely polished with felt-wheels. It is used for beads for necklaces and bracelets, for cabochons, ornamental objects, and sculptures. Branch-like pieces are pierced transversely and strung as spiky necklaces [no. 13]. Corals are sensitive to heat, acids, and hot solutions. The color can fade when worn.

Possibilities for Confusion   With conch pearl (page 232), carnelian (page 126), rhodonite (page 168), and spessartite (page 104). Imitations are made from glass, horn, rubber (gutta-percha), bone, and plastics.

Noble coral (Corallium rubrum)   Most desired of all coral types. According to place of discovery, it has numerous trade names. The color is uniform: light red to salmon-colored (Momo), medium red (Sardegan), ox-blood red (Moro), tender pink with whitish or light-reddish spots (Angel Skin Coral [no. 11]).

Black coral   Consists of an organic horn substance. In the world trade it is, as with blue corals, of no economic importance.

1 Noble coral, 3 beads together 23.77ct	8 White coral, 2 engraved beads
2 Noble coral, branch, Sicily	9 Noble coral, engraved, Italy
3 Noble coral, 2 figurines, Japan	10 Noble coral, branch, Japan
4 Noble coral, 5 cabochons	11 Noble coral, 3 cabochons, 14.05ct
5 Noble coral, 2 necklaces	12 Noble coral, figurine, Italy
6 Noble coral, figurine, Japan	13 White coral, spiky necklace
7 White coral, branches, Japan	14 Black coral, branch, Australia

The illustrations are 50 percent smaller than the originals.

Jet [7, 8]

Color: Deep black, dark brown	Transparency: Opaque
Color of streak: Black-brown	Refractive index: 1.640–1.680
Mohs' hardness: 2½–4	Double refraction: None
Density: 1.19–1.35	Dispersion: None
Cleavage: None	Pleochroism: Absent
Fracture: Conchoidal	Absorption: Cannot be evaluated
Chemical composition: Lignite	Fluorescence: None

Jet (the name comes from a river in Turkey) is a bituminous coal which can be polished. It has a velvety, waxy luster. Deposits are found in England (Whitby), Germany/Württemberg, France (Dép. Aude), Poland, Spain (Asturias), and the United States (Colorado, New Mexico, Utah). It is worked on a lathe. Used for mourning jewelry, rosaries, ornamental objects, and cameos.

Possibilities for Confusion   With anthracite, asphalt, cannel coal (see below), onyx (page 142), schorl (page 110). Some imitations made with glass, rubber (gutta-percha), plastic.

Cannel coal [6]   The name is derived from the English "candle," referring to the wax that was extracted from combustible-rich layers in coal seams; formed predominantly from plant spores and pollen. Deposits are found in Germany, England, and Scotland. Due to its homogeneity and density, it can be easily worked on the lathe; a high luster can be achieved through polishing. Serves as substitute for jet.

Ivory [1–5, 9]

Color: White, creamy	Transparency: Translucent, opaque
Color of streak: White	Refractive index: 1.535–1.570
Mohs' hardness: 2–3	Double refraction: None
Density: 1.7–2.0	Dispersion: None
Cleavage: None	Pleochroism: None
Fracture: Fibrous	Absorption spectrum: Cannot be evaluated
Chemical composition: Calcium phosphate	Fluorescence: Various blues

Ivory originally only referred to the material of the elephant's tusk. Today it is also the teeth of hippopotamus, narwhal, walrus, wild boar, and fossilized mammoth. Most comes from Africa, some from Burma (Myanmar), India, and Indonesia (Sumatra). **Since 1989, there has been a worldwide ban on any trade in elephant's ivory.**

Worked with cutting tools and file, it can be dyed. Used for ornamental objects, pendants, and for costume jewelry. Can be confused with many types of bone [no. 10].

Odontolite (Also called tooth turquoise)   Odontolite (Greek—toothstone) is fossilized tooth or bone substance from extinct prehistoric large animals (such as the mammoth, mastodon, or dinosaurs). Dyed turquoise blue with vivianite (page 208). Deposits are found in Siberia and the south of France. Has become very rare. Can be confused with turquoise (page 170) and ivory that has been dyed blue (see above).

1 Ivory, concentric spheres, China	7 Jet, 3 faceted pieces
2 Ivory, figurine and snuff bottle	8 Jet, 2 cabochons
3 Ivory, rough, Congo	9 Ivory, brooch, China
4 Ivory, necklace, China	10 Bone, dyed, Israel
5 Ivory, bracelet and figurine	
6 Cannel coal, rough and partly polished	
The illustrations are 50 percent smaller than the originals.	

Amber Also called Succinite

Color: Yellow, brown, and also other colors	Transparency: Transparent to opaque
Color of streak: White	Refractive index: 1.539–1.545
Mohs' hardness: 2–2½	Double refraction: None
Density: Mostly 1.05–1.09	Dispersion: None
Cleavage: None	Pleochroism: Absent
Fracture: Conchoidal, brittle	Absorption spectrum: Cannot be evaluated
Crystal system: Amorphous	Fluorescence: Bluish-white to yellow-green
Chemical composition: Approximately	Burmite: blue
$C_{10}H_{16}O$ mixture of various resins	

Amber is the fossilized, hardened resin of the pine tree, *Pinus succinifera*, formed mainly in the Eocene epoch of the Tertiary period, about 50 million years ago; found mostly in the Baltic, although younger ambers are known from the Dominican Republic. Mostly amber is drop or nodular shaped with a homogeneous structure, or has a shell-like formation, often with a weathered crust. Pieces weighing over 22 lb (10 kg) have been found. It is often turbid because of numerous bubbles [no. 8], fine hair lines, or tension fractures. It is possible to clear air bubbles and enclosed liquids from the material by boiling in rape-seed oil. Yellow and brown are predominant colors. There are occasional inclusions of insects or parts of plants (no. 7, and illustration on page 50) and of pyrites.

Amber is sensitive to acids, caustic solutions, and gasoline, as well as alcohol and perfume. Can be ignited by a match, smelling like incense. When rubbed with a cloth, amber becomes electrically charged and can attract small particles. It has a vitreous luster; when it is polished, a resinous luster.

Deposits The largest deposit in the world is west of Kaliningrad, Russia. Under 100 ft (30 m) of sand is a 30 ft (9 m) layer of amber-containing clay, the so-called blue earth. It is mined from the surface with dredging chain buckets. First the amber is washed out, then picked by hand. Only 15 percent is suitable for jewelry. The remainder is used for pressed amber (see Ambroid below) or used for technical purposes.

There are large reserves on the seabed of the Baltic. After heavy storms, amber is found on the beaches and in shallow waters of bordering countries. This sea amber is especially solid and used to be regularly fished for by fishermen. Further deposits are found in Sicily/Italy (called Simetite), Rumania (Rumanite), Burma (Myanmar—Burmite), China, the Dominican Republic, Japan, Canada, Mexico, the United States (Alaska, New Jersey).

Uses Amber has been used since prehistoric times for jewelry and religious objects, accessories for smokers, also as amulets and mascots. The Baltic amber, the "gold of the North," is among the earliest-used gem materials. Used today for ornamental objects, ring stones, pendants, brooches, necklaces, and bracelets.

Possibilities for Confusion With citrine (page 120), fluorite (page 198), meerschaum (page 222), onyx marble (page 218), sphalerite (page 200), ambroid (see below). Imitated by newly created resins (copal), other synthetic resins, and yellow glass.

Ambroid Natural-looking pressed amber made from smaller pieces and the remains of the genuine amber. These bits are welded at 284–482 degrees F (140–250 degrees C) and 3000 atmospheres pressure into a substance that can be easily mistaken for natural amber.

1 Amber, rough	5 Amber, 2 baroque necklaces
2 Amber, partly polished	6 Amber, various colors
3 Amber, 3 cabochons	7 Amber, with insect inclusion
4 Amber, 2 bead necklaces	8 Amber, with bubble inclusion

Pearl

Color: White, pink, silver-, cream-, golden-colored, green, blue, black
Color of streak: White
Mohs' hardness: 2½–4½
Density: 2.60–2.85
Cleavage: None
Fracture: Uneven
Crystal system: (Orthorhombic) microcrystalline
Chemical composition: Calcium carbonate + organic substances + water

Transparency: Translucent to opaque
Refractive index: 1.52–1.66
  Black: 1.53–1.69
Double refraction: –0.156
Dispersion: None
Pleochroism: Absent
Absorption spectrum: Not diagnostic
Fluorescence: Weak, cannot be evaluated
  Genuine black p.: Red to reddish
  River-p.: Strong: pale green

Most pearls are products of bivalve mollusks mainly of the oyster type (family *Ostreidae*). They are built up of mother-of-pearl (nacre), which is mainly calcium carbonate (in the form of aragonite), and an organic horn substance (conchiolin) that binds the microcrystals concentrically around an irritant.

Although the Mohs' hardness is only 2½–4½, pearls are extraordinarily compact, and it is very difficult to crush them.

The derivation of the name pearl is uncertain, but may be from a type of shell (Latin—*perna*) or from its spherical shape (Latin—*sphaerula*).

The size of pearls varies between a pin head and a pigeon's egg. One of the largest fine pearls ever found (called the Hope Pearl after a former owner) is 2 in (5 cm) long and weighs 454ct (1814 grains = 90.8 grams); it is in the South Kensington Museum in London.

The typical pearly luster is produced by the overlapping platelets of aragonite and film of conchiolin nearer to the pearl surface. This formation also causes the interference of light and the resulting iridescent colors (called orient) that can be observed on the pearl surface. The color of pearl varies with the type of mullusk and the water, and is dependent on the color of the upper conchiolin layer. If the conchiolin is irregularly distributed, the pearl becomes spotty.

Formation   Pearls are formed by saltwater oysters (genus *Pinctada*), some freshwater mussels (*Unio*), and more rarely by other shellfish. They are formed as a result of an irritant that has intruded between the shell of the mollusk and the mantle or into the interior of the mantle. The outer skin of this mantle—the epithelium—normally forms the shell by secretion of mother-of-pearl, and also encrusts all foreign bodies within its reach. And such an encrustation will develop into a pearl. If a pearl is formed as a wart-like growth on the inside of the shell, it must be separated from the shell when it is collected. Therefore, its shape is always semi-spherical. It is called blister or shell pearl. In the trade they sometimes cement these to mother-of-pearl backings to form mabé pearls.

1 Mother-of-pearl shell with cultured pearls
2 Pearl necklace, Biwa Lake cultured, baroque
3 Pearl necklace, 4 chokers, silver white
4 Pearl necklace, baroque
5 Pearl necklace, Biwa Lake cultured, graduated
6 4 baroque pearls
7 Pearl, Mabé cultured, silver-white, 20mm
8 Pearl, Mabé cultured, gray, oval
9 6 baroque pearls, 35.71ct
10 10 baroque pearls, Biwa Lake cultured
11 Mother-of-pearl, 2 cut pieces
12 Pearl choker, gray
13 3 pearls, Biwa Lake cultured, 29.67ct
14 4 cultured pearls, 16.16ct
15 Pearl necklace, choker, gray
16 6 black baroque pearls
17 Pearl necklace, black baroque
18 6 gray pearls, 17.28ct
The illustrations are 60 percent smaller than the originals.

When a foreign body enters the inner part of the mantle—the connective tissue—the mollusk forms a nonattached, rounded pearl as a type of immunity defense. The epithelial tissue, which has been drawn into the connective tissue together with the foreign body, forms a pearl sac around the intruder and isolates it by secretion of nacre. As we now know, the nacre can also produce a pearl without any foreign body. It is sufficient that a part of the epithelium may for any reason (for instance, an injury from the outside) be drawn into the connective tissue of the mantle.

Natural Pearls

Natural pearls are those pearls that come into being without intervention by human beings, in the ocean as well as in freshwater.

Sea Pearls   Pearl-producing sea mollusks live along long stretches of coast at a depth of about 50-65 ft (15-20 m). The various species range in size from about 2½ to 12 in (6 to 30 cm); their life span is about 13 years.

Their habitat is the warmer regions on both sides of the equator. The most important occurrences yielding the best qualities (rose and creamy white) have long been those in the vicinity of the Persian Gulf. Because of this occurrence, all natural seawater pearls, wherever they come from, have been called "oriental pearls" in the trade. US FTC guidelines now restrict this term to those "of the distinctive type and appearance of pearls obtained from mollusks inhabiting the Persian Gulf."

Also, in the Gulf of Manaar (between India and Sri Lanka) there are ancient beds (pearls of pink-red and soft yellow color), but the pearls are mostly small (so-called seed pearls). Other important occurrences are along the coasts of Madagascar, Burma (Myanmar), the Philippines, many islands in the South Pacific, northern Australia, and the coastal lines of Central America and northern South America. In Japan, the most important country of cultured pearl production, there are only some small beds with natural pearls. Pearls are harvested by divers. Formerly the work was done without any special tools; today the most modern diving gear is sometimes used. Only every 30th or 40th oyster contains a pearl. In Sri Lanka in 1958, dragnets were experimentally used; the result was catastrophic, as the next growth was almost completely destroyed.

The giant conch (*Strombus gigas*) is a sea snail that produces pearls. Its product (called conch or pink pearl) has a silky sheen that resembles porcelain. Commercially it is not important, and most gemologists do not consider it a true pearl because it lacks nacre.

Cross section of pearl formation in the mantle area of a pearl mollusk.

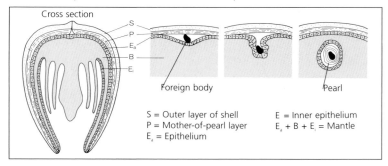

Cross section

S = Outer layer of shell
P = Mother-of-pearl layer
E$_a$ = Epithelium

E = Inner epithelium
E$_a$ + B + E$_i$ = Mantle

Foreign body

Pearl

Very skilled hands are required for the setting of a bead into the pearl mollusk.

River Pearls　Fishing for natural pearls in freshwater, the river pearls, is today of no great importance commercially; they are rarely of good quality. During the Middle Ages and just after, fishing for pearls in the rivers, which are low in lime and rich in oxygen, of Central Europe as well as in rivers of similar habitats in Asia and North America was of some importance.

In Europe, fishing for pearls was the absolute privilege of princes; fished pearls had to be delivered to the ruler personally. Because of pollution of the water, pearl mussels have not flourished or become extinct. Even though the supply in some rivers has partially regenerated due to the improvement of water quality, their existence continues to be threatened because of elevated nitrate levels in the water. In the Scandinavian countries and in Central Europe, pearl mussels are environmentally protected; *pearl fishing is forbidden*. A limited number of natural freshwater pearls is still obtained from rivers in the United States.

Cultured Pearls

In the 20th century depletion and pollution have drastically reduced the supply of natural pearls. The increased demand for pearls has led to their cultivation in large quantities. Such cultured pearls are not an imitation, but a product which has been produced with human assistance. Today cultured pearls amount to 90 percent of the total pearl trade. There are cultured-pearl farms in the ocean as well as in freshwater rivers.

Saltwater cultured pearls　The principle behind pearl culturing is simple. Humans cause the mollusk to produce a pearl by insertion of a foreign body (compare formation of the pearl, page 230). In China as early as the 13th century, small objects were fixed to the inner wall of a mollusk shell so that they would be covered with pearl material. Round pearls were first produced, it is thought, by the Swedish naturalist Carl von Linné (Linnaeus), in 1761 in river mussels.

Modern cultivation of round pearls is based on the experimental work of the German zoologist F. Alverdes, as well as the Japanese, T. Nishikawa, O. Kuwabara, T. Mise, and K. Mikimoto, in the late 1800s and early decades of this century. To stimulate the mollusks to produce pearls, rounded mother-of-pearl beads from the shell of the North American freshwater mussel are at first wrapped with a piece of tissue from the mantle of a pearl mollusk (*Pinctada martensii*), and are then inserted into the cell lining of the mantle of another pearl mollusk.

233

The inserted tissue continues to grow and has the effect of a pearl sac in which pearl material is secreted. The most important element in the production of a pearl is the tissue, not the foreign body. It has been proven that one can do without the bead, but then the process will not remain economical since the culture of a large pearl takes too much time.

Only one layer of nacre is necessary for the bead to acquire the characteristic pearly luster. Since 1976, coreless pearls from the Pacific have been on the market; they are called *Keshi pearls* in the trade. These grow spontaneously (without nucleation) in mollusks previously used for culturing. It has been an ongoing technical dispute whether these are natural pearls, as the people who offer them claim, or whether they are correctly considered cultured products.

The insertion of the bead into the mollusk requires agile, skilled hands; commonly, this job is filled by women. They operate on 300 to 1000 oysters a day. The normal size of a bead (0.24–0.27 in/6–7 mm) requires a three-year-old mollusk; for smaller beads, younger mollusks can be used. When the bead is greater then 0.35 in (9 mm), the mortality rate of the oysters rises to 80 percent.

Prepared mollusks are kept underwater in the bay in wire cages or, preferably nowadays, plastic cages suspended at a depth of 6½–20 ft (2–6 m) and hung from bamboo floats or ropes which are fixed to buoys. Several times a year the mollusks and their cages must be cleaned and freed from seaweed and other deposits. Their natural enemies are fishes, crabs, polyps, and various parasites—mainly a zooplankton which appears in large quantities (a "red tide") and endangers a whole cultivation farm because it consumes large quantities of oxygen.

Mollusk cages must always be supervised and several times a year cleaned of seaweed and other unwanted deposits.

Rafts of cultivated pearl farms in an ocean bay in southern Japan.

The temperature of the water has a great influence on the growth of the mollusks; the Japanese variety dies at 51 degrees F (11 degrees C). If the temperature suddenly falls before the winter, the floats with the submerged burden must be dragged from northern farms to warmer waters. The yearly growth rate of the pearl layer surrounding the bead in Japan used to be 0.09 mm; it has now reached 0.3 mm and is said to be 1.5 mm in southern seas.

Some cultivation farms have been transferred from bays to the open sea, as it was thought that the water current would activate the mollusks to produce faster, with better shapes. At the same time, the bays with their innumerable floats would be less crowded and conditions improved for other pearl mollusks.

The mollusks remain in the water about two to three years; by then, the nacre layers around the bead are about 0.5–1.0 mm. If they remain longer in the water, there is a danger that they will become ill, die, or mar the shape of the pearl. No mother-of-pearl is secreted after the 7th year.

A saltwater mollusk can usually be used only once. After the pearl has been removed, most of them die. Accordingly, it is important to make sure that there is sufficient aftergrowth. Cultured pearls with a very thin nacre shell are considered to be of inferior quality.

The best times for harvesting in Japan are the dry winter months, November to January, as the secretion of the mother-of-pearl is halted and an especially good luster is present. The pearls are taken from the mollusks, washed, dried, and sorted according to color, size, and quality. The whole production yields roughly 10 percent for good-quality jewelry; 60 percent are of minor quality, and 15-20 percent are rejected.

In order to improve and/or change the colors, cultured pearls are treated in various ways, such as bleaching, dyeing, or with radiation. Occasionally, the color is also influenced by insertion of colored cores. Colors that are achieved through radiation are not always permanent.

The first Japanese farms were founded in 1913 in southern Honshu; today there are also some on Shikoku and Kyushu. Since 1956, pearls of large size and good quality have been grown in coastal waters of northern and western Australia, as well as cultured blister pearls with a diameter of 0.6–1 in (15–22 mm), called Mabé pearls in the trade. Today, numerous farms exist in South-Southeast Asia, among others in southern Burma (Myanmar), Malaysia, and Indonesia. Among the islands of French Polynesia, the famous black Tahitian pearls are cultured.

Cultured Freshwater Pearls   Since the 1950s there has been a freshwater pearl farm in the Biwa Lake (Japanese—*biwako*), north of Kyoto on Honshu, Japan. Pieces of tissue measuring 4 × 4 mm, usually without solid beads, are inserted into the fresh water mussels (*Hyriopsis schlegeli*). As these mussels are very large (8 × 4.3 in/20 × 11 cm), ten insertions can be made into each half and, sometimes, an additional one with a mother-of-pearl bead. For each insertion a sac with pearl is formed. After one to two years the pearls are already 0.24–0.28 in (6–8 mm) large, but rarely round. Therefore, they are taken out of the mussel, covered with new tissue, and inserted into the same or another mussel to improve the shape. Biwa cultured pearls reach a diameter of 0.5 in (12 mm), but rarely have a perfect round shape. "Rice" and button shapes are most common.

The life of a freshwater mussel is 13 years, but after the operation it produces mother-of-pearl only for three more years. Many mussels can be harvested three times.

Cultivation methods are the same as for the sea mollusks. Cages are hung on a bamboo frame at about 3–6½ ft (1–2 m) depth. The success rate is about 60 percent, which is clearly higher than in seawater, probably because there are fewer dangers in Biwa Lake.

Since the beginning of the 1970s, freshwater pearls are also cultivated in China. They are brought onto the market now in large quantities. But their quality is not as good as that of the Japanese cultured pearls. Since the early 1990s, round bead-nucleated Chinese freshwater cultured pearls have been increasingly available.

A skilled eye is necessary for the sorting and quality evaluation of pearls.

Pearls are drilled in such a way that damaged or less perfect spots "disappear" at the same time.

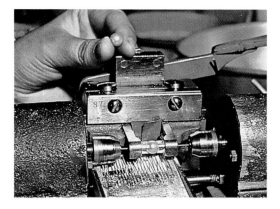

Use and Valuation of Pearls

Pearls have been regarded as one of the most valuable gem materials. They have been used for adornment for 6,000 years. In 2500 B.C., there was a substantial pearl trade in China. Pearls are also popular because they do not require any processing; in their natural state they show their full gloss, the desired luster.

As much as 70 percent of all pearls are strung and worn as necklaces. In the United States, the most popular length, known as the "Princess," is about 17 to 19 in (43 to 48 cm). If pearls of the same size are used, one speaks of a uniform necklace; if they are graduated in size with the biggest in the middle and the smallest at the ends, one speaks of the graduated necklace. The combination, i.e., the careful selection of the pearls for necklaces or collars, is done by eye.

Uses A point of the pearl, which either has a mark or is less perfect, is chosen for drilling a hole, thus eliminating the mark. The diameter of the drill hole, according to international agreement, should be 0.3 mm. To fix the pin for earrings, brooches, pins, and rings, a drill hole to the depth of two-thirds or three-quarters of the pearl's diameter should suffice. Blue pearls should never be drilled, as they may change color when air reaches the drill hole.

Spotted or damaged pearls can be peeled, i.e., the outer layer can be removed. Badly damaged parts can be cut away; the remaining part is traded as half or three-quarter pearl (not to be confused with blister pearls). They are used mainly for earrings and brooches.

For decades, the United States has been the largest buyer of cultured pearls.

Valuation The pearl is valued according to shape, color, size, surface condition, and luster. The most valuable is the spherical shape. Those flattened on one side or half-spherical pearls are called bouton (French—button) or button pearls; irregular pearls are baroque pearls. Pearls long worn in a necklace take on the shape of a little barrel; one speaks of barrel pearls.

Fair-skinned women in Europe and the United States prefer white or rose color; dark-haired ladies prefer cream-colored pearls.

Natural pearls are weighed in grains (1 grain 0.05 g = 0.24ct or ¼ carat), and today increasingly in carats. The Japanese weight of *momme* (= 3.75 g = 18.75ct) is becoming rarer in European trade circles. Traditionally, natural pearl prices were calculated according to weight. Except in large wholesale transactions, cultured pearl prices are normally quoted with regard to size.

237

The word *pearl* without addition should be used only for natural pearls. Cultured pearls must be designated as such.

Taking Care of Pearls

Since conchiolin is an organic substance, it is prone to change, especially to drying out. This can lead to an "aging" of the pearls, limiting their useful life. At first they became dull, then fissures occur, and finally the beads spall. There is no guarantee possible for the life span of a pearl; on average, one estimates 100 to 150 years. But there are pearls which are some hundreds of years old and still look their best. Proper care can certainly preserve and extend the life of pearls. Extreme dryness is damaging; pearls are also sensitive to acids, perspiration, cosmetics, and hair spray.

Regular examination and maintenance by a firm specializing in pearls can also prolong their life.

Since pearls have a low hardness, they can be easily scratched. Therefore, wear and store them in such a way that the pearl surface is never in contact with metal or other gemstones.

Discerning Pearls from Imitations

As is the case for all gemstones, there are numerous imitations of pearls on the market. To recognize these is just as important as the differentiation between natural and cultured pearls, because there is a huge disparity in respective prices.

Differentiating between Natural and Cultured Pearls   There is little or no difference in the appearance of natural and cultured pearls. To differentiate between them is difficult. Their density can help, as it is greater than 2.73 in the case of most (but not all) cultured pearls, whereas the density of natural pearls is usually lower. Sometimes an examination with certain radiation can clarify. Under ultraviolet light, for instance, cultured pearls have a yellow luminescence, and under X rays a green one, but these reactions are not completely dependable. One reliable method of differentiating between cultured and natural pearls is by examining their inner structure. Natural pearls have a concentrically layered structure, whereas the inner structure of the cultured pearl varies according to the type of bead. Typically, an expert uses an instrument like an endoscope to explore the structure of the pearl inside the drilling hole.

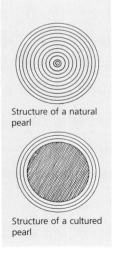

Structure of a natural pearl

Structure of a cultured pearl

Natural and cultured pearls distinguish themselves through their inner structure.

Radiography methods with X rays, such as the X ray diffraction (lauegram) method and the X ray silhouette (shadowgraph) procedure, are effective. They can be applied to both drilled and undrilled pearls. In cultured pearls, they detect the bead structure or the thickness of the "natural" pearl layer around the foreign body.

Imitations   A fine imitation is the so-called fish-scale pearl. It consists of glass or enamel coated with *essence d'orient*, which is produced from the scales of certain fish. In other imitations, part of a sea snail (antilles pearls), mussels (takara pearls from Japan), or teeth (of the sea cow-dugong pearl) are used. There are also plastic products on the market. Under FTC guidelines, all of these must be clearly identified as imitations.

The mabé pearl (sometimes called Japanese pearl in the trade) may also be included among the imitations, because this is not a cultured pearl in the trade sense. It consists of a thin mother-of-pearl layer and some other artificial part. A clay or resin bead is fixed to the inner part of a shell and is then covered with a thin pearl layer. After the harvest the bead is removed and replaced by a mother-of-pearl half-bead. The nacre shell is then filled with epoxy and cemented to a mother-of-pearl backing.

Operculum   The operculum (misleadingly called Chinese cat's-eye) has a structure similar to a half-pearl, with a porcelain-type coloring, but is actually the slightly arched lid of a sea snail found in the Australasian islands area, where it is used for adornment. It is not well known in Europe or the United States.

Mother-of-Pearl [page 231, nos. 1, 11]

The inner nacreous layer of a mollusk shell, or sometimes of a snail shell which has an iridescent play of color, is called mother-of-pearl ("mother of the pearl").

Mother-of-Pearl of the Pearl Mollusk   The mother-of-pearl of the pearl mollusk is most often used. Accordingly, the main suppliers are the pearl farms. The basic color is usually white; it is naturally dark in the mother-of-pearl from Tahiti.

For structure, formation, and distribution see the discussion of pearls starting on page 230. Mother-of-pearl is favorably used for ornamental purposes, for clock faces, buttons, costume jewelry, and for inlaid work (such as the handles of knives and pistols).

Mother-of-Pearl of the Paua (abalone)   The mother-of-pearl of the Paua (*Haliotis australis*) from New Zealand, which has a blue-green iridescent color play, has been used by the indigenous Maori people for centuries for inlays in mystical carvings. Similar animals found along the U.S. coasts of Florida and California are called abalone.

The mother-of-pearl of this mollusk has now been used for some time in the Western world, especially for costume jewelry. It is called sea opal, because of a resemblance to the color effects seen in opal.

Mother-of-pearl of the Paua with striking iridescence.

New on the Market

By highlighting gemstones described as being "new on the market," it does not necessarily mean that these stones were found only recently. The particular mineral or aggregate may very well have been known for a long time, but its importance for the gemstone trade has remained minor or insignificant so far due to its extreme rarity of occurrence in gemstone quality or because it has been poorly marketed.

1 **Verdite** [2 cabochons]   Light to dark green, often spotty serpentine rock. Named for its green color. Mohs' hardness about 3, density 2.8–3.0. Translucent to opaque. Used for sculptures; increasingly also for costume jewelry. Has been used for cultural purposes for a long time in southern Africa.

2 **Charoite** [3 cabochons]   Lilac-colored to violet. Through admixture of accompanying minerals, it becomes white- or black-spotted or flamed. Translucent to opaque. Mohs' hardness 5–6. Density 2.54–2.78. Monoclinic, $K(Ca,Na)_3Si_4O_{10}(OH,F) \cdot H_2O$. Vitreous to silky luster. Named after river in Siberia, Russia. Was recognized in 1978 as independent mineral.

3 **Gneiss**   Opaque metamorphic rock with structure that produces a decorative effect. Main parts of the aggregate are feldspars and quartz. Compact and hard. Density 2.65–3.05. Besides gray and greenish, also brownish and reddish. Used for costume jewelry and ornamental objects.

4 **Unakite**   Opaque granitic rock with main parts of aggregate being quartz and feldspar as well as greenish epidote. Very compact and hard. Density 2.85–3.20. Named after place of discovery in South Carolina, United States. En cabochon and barrel-shaped for costume jewelry and ornamental objects.

5 **Nuummite** [2 cabochons]   Opaque rock of an almost black basic color with main parts of aggregate being gedrite and anthophyllite; other parts are pyrite, magnetite, and chalcopyrite. Because of a lamellar, fibrous structure, there is an iridescent play of colors. Mohs' hardness 5½–6. Density around 3. Named after place of discovery in Greenland. It is worked flat or en cabochon.

6 **Ammonite** (Also called ammolite and korite) [3 flat cuts]   Fossilized shell of ammonites in the form of aragonite or calcite (page 208). Because of a lamellar structure of the glass-like shell, there is an iridescent play of colors. Mohs' hardness 4. Density 2.75–2.80. Place of discovery in Alberta, Canada; on the market since 1969. Because it is not very thick, it is often worked into doublets or triplets.

7 **Carletonite** [3 faceted stones]   Deep blue, also pale blue. Transparent. Mohs' hardness 4–4½. Density 2.45. Tetragonal. Vitreous luster.

8 **Catapleiite** [3 faceted stones]   Colorless. Transparent. Mohs' hardness 5–6. Density 2.72. Hexagonal, $Na_2ZrSi_3O_9 \cdot 2H_2O$. Vitreous luster.

9 and 11 **Sugilite** [2 cabochons, 2 cabochons]   Violet. Translucent to opaque. Mohs' hardness 6–6½. Density 2.76–2.80. Hexagonal, $KNa_2(Fe,Mn,Al)_2Li_3Si_{12}O_{30}$. Resinous luster. Formerly erroneously offered as Sogdianite.

10 **Gaspeite** [2 cabochons]   Light green. Translucent to opaque. Mohs' hardness 4½–5. Density 3.7. Trigonal, $(Ni,Mg,Fe)CO_3$. Vitreous luster.

The illustrations are 10 percent larger than the originals.

Imitation and Synthetic Gemstones

There is nothing against the law in producing imitation or synthetic gemstones, as long as no one is harmed or defrauded by it. These products are indeed an important element of the gemstone trade. Those who do not want the security risk of, or cannot afford, genuine gemstones can use these for their adornment. But when imitations or syntheses are passed off as more valuable true gemstones at inflated prices, then that is fraud. Imitations and syntheses must always be designated as such; they have to be named correctly. A distinction is also made between imitations, which only look similar to the gems, and syntheses, which are artificial versions of natural gemstones (compare the definitions on pages 10 and 11).

Imitations

Ancient Egyptians were the first who feigned gemstones with glass and glaze, because genuine were too expensive and/or too rare.

In 1758, a Viennese, Joseph Strasser, developed a type of glass which served for a long time as a substitute for diamond. It could be cut and was, in fact, very similar to diamond in appearance, due to its high refractive index. Even though production and sale was prohibited by the Empress Maria Theresa, this diamond imitation, called *strass*, reached the European trade via Paris.

Until 1945, Gablonz and Turnau in Czechoslovakia were important centers for the glass-jewelry industry. Then this tradition was taken over by Neugablonz in the Allgäu, Bavaria. For costume jewelry, cheap glass was used. For more valuable gemstone imitations, lead or flint glass with a high refractive index is used. Porcelain, enamel, and resins as well as other plastics also serve as gemstone imitations. Most imitations have only a color similar to that of the gemstone; other properties, such as hardness or fire, could not be satisfactorily imitated.

New Imitations A special category of imitation is the particular synthetic stone, which, even though it does not have a counterpart in nature, is very similar to other gemstones in its physical properties and especially good optical effects. These gem-quality substances, with no analogous natural gemstone, are nonetheless counted among imitation gemstones. They usually serve as a diamond substitute.

To this category belong synthetic rutile (TiO_2; trade names include titania and diamonite), strontium titanate ($SrTiO_3$; fabulite, diagem), YAG (yttrium aluminum garnet, $Y_3Al_5O_{12}$; diamonaire, diamonique), and GGG (gadolinium gallium garnet,

A selection of gemstone doublets and gemstone triplets

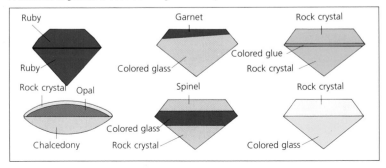

The synthetically made gemstone YAG is a good diamond imitation.

Gd$_3$Ga$_5$O$_{12}$; 3G, galliant). Since the 1970s synthetic cubic zirconia (ZrO$_2$; CZ, djevalite, phianite, zirconia) has been the most popular diamond imitation, but in recent years synthetic moissanite (SiC; silicon carbide sold as C3) has appeared on the market and is considered the best yet by many experts.

Altogether, the list of artificial products that serve as diamond substitutes includes more than a dozen materials marketed under scores of different (often confusing) trade names around the world.

Combined Stones

A popular type of the gemstone imitations are fabrications called combined or assembled stones, in which one part may be a genuine gem, another part glass, foil, or plastic. There are many combinations; sometimes two natural gemstone pieces are made even more beautiful with a colorful glue layer to make a larger stone. Stones with two parts are called doublets; stones with three parts, triplets (see the illustration on page 242).

Stones which are carefully constructed are difficult to recognize, especially when the seams are covered by the setting.

Synthetic Gemstones

The dream of mankind to produce artificial stones that are really the same as the natural gemstones was realized at the end of the 19th century. The French chemist A. V. Verneuil succeeded around 1888 in synthesizing rubies at a commercial price. In fact, 50 years earlier the first gemstones had been produced synthetically, but they were only of scientific interest.

The flame fusion process (see page 244) developed by Verneuil is still largely used today. The method is as follows: powdered raw material (aluminum oxide with dyeing additives) is dropped through a high-temperature oxyhydrogen flame that melts the powder. The molten drops fall on a cradle, where they crystallize and form a pear-shaped "boule" (see page 245). Although this boule does not have any recognizable crystal faces, the inner structure is the same as that of a natural crystal. The boules grow to about 3 in (8 cm) thick and several inches high. The growth time is several hours.

Verneuil first produced rubies, followed in 1910 by synthetic sapphires; later on, colorless, yellow, green, and alexandrite-colored corundums were produced. By adding rutile components to the smelting, the cultivation of synthetic star rubies and star sapphires succeeded in the United States in 1947, following Verneuil's method.

243

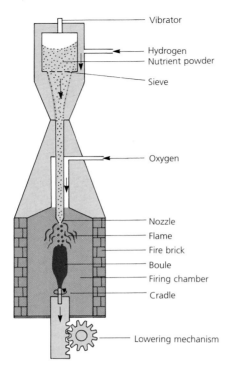

Vibrator
Hydrogen
Nutrient powder
Sieve
Oxygen
Nozzle
Flame
Fire brick
Boule
Firing chamber
Cradle
Lowering mechanism

Gemstone Synthesis Following the Flame Fusion Process (after Verneuil) Powdered raw material melts in an oxyhydrogen gas flame and drips onto a cradle, where a boule builds up. By the same degree that the bulb grows upward, the cradle is lowered, so that the upper surface of the boule is always the same distance from the burner nozzle.

Picture right: Melting boules and synthetic gemstones which have been cut out of them.

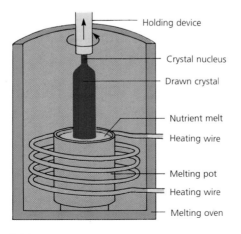

Holding device
Crystal nucleus
Drawn crystal
Nutrient melt
Heating wire
Melting pot
Heating wire
Melting oven

Gemstone Synthesis Following the Drawing Procedure (after Czochralski) A boule is, so to speak, drawn out of the melt, after a crystal nucleus has initiated the growth of the boule at the surface of the melt. Under rotation, the forming boule is continually drawn upward, while it grows respectively on the underside.

Following the Verneuil process, synthetic spinels have also been produced since 1910. Their composition, though, is somewhat different from that of natural spinels. With the addition of heavy metals, very good color tints of other gemstones can be achieved, for instance, those of aquamarine and of tourmaline.

Synthetic emeralds of usable gem size have existed only since the 1940s, although experiments had been conducted for over a hundred years.

Large synthetic crystals of highest purity can also be cultivated with the drawing procedure that was developed in 1918 by the German chemist I. Czochralski. The cultivation product is drawn out of the nutrient melt, after a crystal nucleus has initiated the growth of the boule (see the illustration on page 244).

In the early 1950s diamond synthesis succeeded in both Sweden and the United States. The initial product was not gem quality, but industrial-grade synthetic diamonds have since become very important as abrasives. Gem-quality synthetic diamonds were first produced in 1970, and they have also found essential applications in science and industry. In recent years gem synthetic diamonds have begun to appear in the jewelry market, and many experts believe they will become increasingly available in the future.

Today, there are hardly any gemstones left for which a synthetic version has not or could not be produced. But whether each possible synthesis can succeed on the market is more a question of the relationship between cost of production and market value. Over a dozen different synthesis procedures are known. Certainly there are others because companies generally regard these procedures as proprietary information and so keep their advances and details secret.

Reconstructed Gemstones   The so-called reconstructed or recrystallized stones are often counted as synthetics. These are gemstones for which little splinters and/or powder of genuine gemstones is melted, sintered, or pressed to form larger pieces. Products made in such a way are widespread, especially for amber, hematite, coral, lapis lazuli, malachite, and turquoise.

Genuineness Test for Diamond

It is often very difficult to distinguish genuine gemstones from synthetics, imitations, or some other fabrications. The refinement and delicacy of execution of gemstone imitations are becoming more and more perfect. Often, a test for genuineness is indicated only after very close examination.

A testing device that is easy to handle, and that in many cases results quickly in a reliable answer, does exist for diamonds. The tester (a heat resistance tester) uses the different abilities of diamond and stones that are diamond substitutes to conduct heat. A special advantage of the device is that it can also be used for stones that are already set. Even the smallest pieces can be tested with its fine metal tip.

Heat resistance tester to identify diamond imitations.

Symbolic and Beneficial Stones

Gemstones represent something special due to their color, luster and form, but also due to the rarity of them. They have therefore always been surrounded by a touch of the mysterious. It is assumed that they possess powers protecting against injury from the outside, or for gaining inner strength, but especially also for medicinal purposes.

Cosmic-Astral Symbolic Stones

Gemstones are awarded symbolic meaning in several respects. Some states, for example, identify themselves with a precious stone mined in their country. Sometimes, precious stones are a symbol for power, status, and wealth. Frequently they are an amalgamation of a wishful imagination of supernatural powers and effective magic.

In relation to mystical notions about connections between man, earth, and cosmos, precious stones become a symbol for fascination and magic, and also become amulet and talisman.

Gemstones as Amulets and Talismans

All primitive races practice in some way warding off evil forces of nature and court the good spirits to be well disposed to them. With increasing cultural progression, precious stones are gaining a growing importance as amulets and talismans.

Amulets are worn on the body. They are to stave off harm, to repel the "evil eye." Sometimes, they are formed in a defensive posture with a finger that is straight or in any other way conspicuously pointed. Besides metal and ceramic, primarily coral serves as the material.

Raw and cut gemstones and pearls are also made into such amulets that are worn as protection, for example, on a neck chain, the arm, or the ear.

Devotionals and rosaries, in the last century often artfully produced from cut precious stones, are amulets blessed by the Church.

Talismans are not worn on the body, but kept in front of the house, in the apartment, or taken along in the car. They are to guarantee health, bring good luck, and be helpful in other ways.

Many gemstones, especially the larger ones such as agate aggregates, mineraloids, drusies, and geodes are used as talismans. The scarabaeus, once regarded as holy in Egypt, is also nowadays cut as a talisman out of various gemstones.

Amulet made out of red coral.
17th cent., length 5 cm.
Treasury of the Residence Munich.

Planet Stones

In Antiquity, and also in the Middle Ages, the formation of gemstones was seen as an efflux of the stars. There are many colorful precious stones that glisten and sparkle in the same way as the stars emit light and colors. The cosmos reflects, so to speak, in the gemstones.

To every planet a gemstone is assigned magic qualities. No assignment system can be detected. Each historian of Antiquity reports different planet stones.

At the present, the planet stones are undergoing a renaissance, with numerous additions in the list of offerings.

Assignment of Gemstones to the Planets in Modern Literature

	Uyldert, 1983	Raphaell, 1987	Richardson/Huett, 1989	Ahlborn, 1996
Mercury	Citrine yellow Sapphire Nephrite Topaz	Chrysocolla Turquoise	Garnet	Chrysolite, Heliotrope, Tiger's-eye
Venus	Nephrite Topaz Rose quartz blue Sapphire Emerald	Kunzite, rose Tourmaline	Chrysoberyl Malachite, Moon-stone, Pearl, Sapphire, Topaz	Malachite, Turquoise
Mars	Garnet, Ruby, Carnelian, Silex	Bloodstone, Carnelian	Bloodstone, Jasper, Onyx, Sardonix	Heliotrope, Rhodochrosite, Rhodonite, Tiger's-eye
Jupiter	Hyacinth, orange Carnelian	Azurite, Lapis lazuli	Jade, Turquoise	Citrine, Cross Stone Sardonyx
Saturn	Gagate, Onyx, Chalcedony Spinel	Malachite, Peridot green Tour-maline	Quartz, Tiger's-eye	
Uranus	Amazonite, Malachite, Turquoise	Aquamarine, Chrysoberyl, Coelestine		
Neptune	Amethyst, Opal	Amethyst, Fluorite	Aquamarine, Azurite, Diamond, Coral, Moonstone, Opal Quartz, Spinel, Tourmaline	
Pluto	Almandine, Bloodstone, Pyrope	Garnet, Ob-sidian, Smoky quartz, Ruby	Amethyst, Jade, Kunzite, Spinel, Zircon	

Zodiac Stones

Similarly to the planet stones, there existed in Antiquity belief in the interrelation of certain constellations, the so-called zodiac signs, to man and gemstones.

These ideas were further nourished in the Middle Ages, and with many people they are nowadays equally topical. Jewelers and dealers in precious stones add to these notions in order to advance their business.

Here too, there is no recognizable coordination system of the gemstones, nor did one ever exist. The gemstones of the zodiac signs listed on the following page were taken from contemporary literature.

Zodiac Stones

Zodiac signs	Illustration		Additional Zodiac Stones
Aries 3/21–4/20	red Jasper, Carnelian		Bloodstone, Chalcedony, Chrysoprase, Ruby, Silex
Taurus 4/21–5/20	Carnelian, Rose quartz		golden Topaz, Coral, Lapis lazuli, Quartz, Sapphire, Sard, Emerald
Gemini 5/21–6/21	Citrine, Tiger's-eye		Agate, Aquamarine, Rock crystal Chrysocolla, Jasper, Onyx, Topaz, Turquoise
Cancer 6/22–7/22	Chrysoprase, Aventurine		white Chalcedony, Chrysolite, Diamond, Carnelian, Moonstone, Rhodochrosite, Emerald
Leo 7/23–8/23	Rock crystal Goldquartz		Almandine, Amber, Chrysolite, Citrine, Diamond, Carnelian, Onyx, Ruby, Sulfur
Virgo 8/24–9/23	Citrine, yellow Agate		Amazonite, Beryl, Jasper, Carnelian, Sardonyx, Turquoise, Zircon
Libra 9/24–10/23	orange Citrine, Smoky quartz		Aventurine, Beryl, Diamond, Jade, Cunzite, Nephrite, Opal, Peridot, Sardonyx, Emerald, Topaz, rose Tourmaline
Scorpio 10/24–11/22	deep red Carnelian		Agate, Aquamarine, Chalcedony, Chrysoprase, Garnet, Obsidian, Smoky quartz, Ruby, Topaz
Sagittarius 11/23–12/21	Sapphire, Chalcedony		Amethyst, Rock crystal, Beryl, Garnet, Pyrope, Sapphire quartz, Sodalite, Spinel, Topaz
Capricorn 12/22–1/20	Onyx, Quartz– Cat's-Eye		Amethyst, Beryl, Gagate, Malachite, Obsidian, Peridot, Smoky quartz, Rose quartz, Ruby, green Tourmaline
Aquarius 1/21–2/19	Turquoise, Hawk's-eye		Amazonite, Amethyst, Aquamarine, Chrysocolla, Coelestine, Garnet, Jasper, Malachite, Obsidian, blue Sapphire
Pisces 2/20–3/20	Amethyst Amethyst Quartz		Aquamarine, Blue-quartz, Diamond Jade, Moonstone, Opal, Sapphire, Sugilite

Gemstones of the Months

Gemstones of the Months

January
Garnet
Rose quartz

February
Amethyst
Onyx

March
Tourmaline
Blood Jasper

April
Sapphire
Diamond
Rock crystal

May
Emerald
Chrysoprase

June
Pearl
Moonstone

July
Ruby
Carnelian

August
Onyx
Sardonyx

September
Peridot

October
Aquamarine
Opal

November
Topaz
Tiger's-eye

December
Zircon
Turquoise

Gemstones of the Months

Originally, the respective zodiac signs were regarded also as the birthstones. They were either amulets or talismans. Only in more recent times, gemstones of the months are according to the calendar also propagate as birthstones, aside from the zodiac gemstones. Here again, arbitrarily assembled gemstone groups are supposed to unfold magic effects. Whenever people feel called upon to convey their own mythological-astrological interpretations, various gemstones are always named as indicators of magic effect. The German poet Theodore Koerner (1791–1813) names in his poem "The Stones of the Months" the following assignments of gemstones to the months:

January	Hyacinth
February	Amethyst
March	Heliotrope
April	Sapphire
May	Emerald
June	Chalcedony
July	Carnelian
August	Onyx
September	Chrysolite
October	Aquamarine
November	Topaz
December	Chrysoprase

Currently, gemstones are also offered for the times of the year, for spring, summer, fall, and winter, and even for every day of the week.

In a publication of 1985, the following recommendation for gemstones for the days of the week can be found:

Sunday	Amber, gold Topaz
Monday	Moonstone, Pearl
Tuesday	Ruby, Garnet
Wednesday	Turquoise, Sapphire, Lapis lazuli
Thursday	Amethyst
Friday	Emerald, Malachite
Saturday	Diamond

Medicinal Stones

Just as the gemstones are regarded as symbols for a supernatural link of man to the sun, the moon, and the stars, they also symbolize magic and curative science.

Powers attributed to the gemstones can prevent or heal illnesses and mitigate or eliminate other infirmities. No authentication or scientific proof exists for the curative powers of gemstones.

Historical Survey

Written accounts about healing or the prevention of illnesses with gemstones have been extant since Antiquity. Eminent writers of the old Age, such as Aristotle, Gajus Plinius Secundus, Dioscurides, and later Marbod, Albertus Magnus, and Konrad von Megenberg, wrote about it.

Under the influence of the Church, the curative effects of the gemstones described in the Antiquity were in part given new meanings to eliminate influences of paganism, and to lead the faithful to an ethical mode of life.

Over centuries, the healing prescriptions by the Abbess Hildegard von Bingen (1098–1179), written down in her book *Physica*, found great acceptance. This gemstone medicinal science (lithotherapy) is experiencing a resurrection. Whole bookstore shelves are filled with discourses about the "Hildegard medicine."

Since according to Hildegard's views, gemstones form through the combined actions of fire and water, they possess powers corresponding to those natural phenomena. In addition, there is God's influence, which wants the gemstones to be seen as a blessing for that which is honorable and useful.

Hildegard discusses in length 24 gemstones and their medicinal effects; some other stones are mentioned in passing.

An example about the sapphire:

Who is dull and would like to be clever, should, in a sober state, frequently lick with the tongue on a sapphire, because the gemstone's warmth and power, combined with the saliva's moisture, will expel the harmful juices that affect the intellect. Thus, the man will attain a good intellect.

Hildegard von Bingen (1098–1179). Statue in the middle-shrine of the Hildegard-Altar, in the Rochus Chapel, Bingen (Germany).

Healing Stones Nowadays

If healing gemstones are mentioned these days, it does not mean that illnesses are cured with gemstones, but that—according to the beliefs of the users—all kinds of negative effects on man can be influenced positively.

Powers and Effects of the Gemstones

All gemstones, a few other stones, and even solidified organogenic (developed from organic materials) products, are used.

The energetic powers ascribed to them by their users stem from Mother Earth during the formation of the stones, or from the sun's warming rays. Therefore, it is recommended to put from time to time healing stones into the sun, or expose them to the (not really warming) moonlight, so that they can replenish the energy that has meanwhile been used. Example: "It is most effective if the stones are put on the terrace or the window sill two nights before full moon."

Another manner of energy-replenishment can be achieved by "burying [the stone] overnight in the ground and afterwards rinsing it for a short time." Synthetic stones may not be used as healing stones. "Only the genuine gemstones grown in nature in the course of millions of years have stored the powers of nature. The artificially produced gemstones lack the soul that has mastery of all."

The much-praised powers of the healing stones have not yet been scientifically proved. All comments such as "proved manifold," "carefully tested," "researched over years," and "scientifically documented," are sheer assertions.

The question as to whether any side elements of the gemstones and the trace elements can influence the human body and the psyche is completely open at the present time. Therefore, traditional medicine rejects a therapy with gemstones.

Drum Stones, i.e., gemstones rounded on all sides. Above, from the left: Amethyst quartz, Rose quartz, Aventurine. Below, from the left: Rhodonite, Larimar, Agate.

Without any doubt, there are certain healing results with lithotherapy (treatment with stones). But that probably is not due to the supposed medicinal powers of the gemstones; it is a so-called placebo effect, an imaginary effect, a success by suggestion. The lithotherapy obviously is in general nothing else but a "psychotherapy."

Preparation and Usage of the Gemstones

In order to maximize the effect of the healing stones, at times mystical and symbolic signs are engraved on the stone, as was practiced in Antiquity and the Middle Ages. The shape of the gemstones may be also be useful in the curative treatment, according to therapists. As a pendulum, for example, rock crystals are used that are formed as symmetrical as possible, but also quartz and other gemstones that have been shaped. A pronounced tip is supposed to increase the success in an individual search for effective stones, and in diagnosing sicknesses of the body.

Very popular are baroque or drum stones rounded on all sides—the so-called fondling-stones that are held in the hand, or carried in a pocket. Yet any other natural or cut form of a gemstone can supposedly be just as helpful.

The use of the gemstones is direct, by skin contact, ground in the form of powder or as a pill, as a means for meditation, or indirect as an essence for curative drinks, and in gemstone elixir for poultices. Stomachaches are helped by ruby-water, and for protection and illumination one is to take diamond- and rock crystal-water (according to an author in 1985).

According to a recent publication, for beneficial results the following is required: "Discharging of the stone before the treatment (by rinsing with cold water, or burying for some days), especially with newly bought, gifted, or inherited stones."

Sometimes, modern literature about beneficial stones is a reflection of the darkest Middle Ages. The borderline between magic, sorcery, fantasy, and therapeutic healing is not recognizable in lithotherapy.

Pendulums with a pronounced tip used in healing with gemstones. From the left: Rock crystal (smoothed), Citrine (untreated), Sugulite (cut), Smoky topaz (smoothed).

Healing with Gemstones

Examples of illnesses that allegedly can be healed. Assembled from current literature (selection).

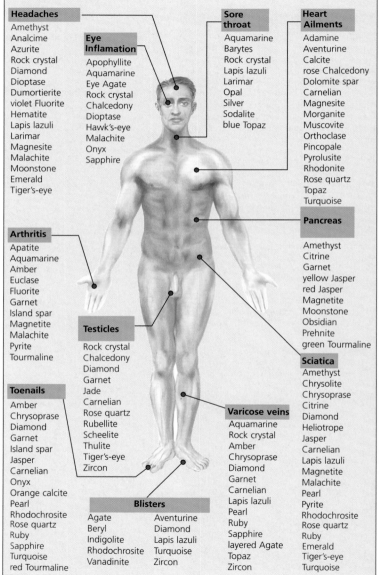

Headaches

Amethyst
Analcime
Azurite
Rock crystal
Diamond
Dioptase
Dumortierite
violet Fluorite
Hematite
Lapis lazuli
Larimar
Magnesite
Malachite
Moonstone
Emerald
Tiger's-eye

Eye Inflamation

Apophyllite
Aquamarine
Eye Agate
Rock crystal
Chalcedony
Dioptase
Hawk's-eye
Malachite
Onyx
Sapphire

Sore throat

Aquamarine
Barytes
Rock crystal
Lapis lazuli
Larimar
Opal
Silver
Sodalite
blue Topaz

Heart Ailments

Adamine
Aventurine
Calcite
rose Chalcedony
Dolomite spar
Carnelian
Magnesite
Morganite
Muscovite
Orthoclase
Pincopale
Pyrolusite
Rhodonite
Rose quartz
Topaz
Turquoise

Pancreas

Amethyst
Citrine
Garnet
yellow Jasper
red Jasper
Magnetite
Moonstone
Obsidian
Prehnite
green Tourmaline

Arthritis

Apatite
Aquamarine
Amber
Euclase
Fluorite
Garnet
Island spar
Magnetite
Malachite
Pyrite
Tourmaline

Testicles

Rock crystal
Chalcedony
Diamond
Garnet
Jade
Carnelian
Rose quartz
Rubellite
Scheelite
Thulite
Tiger's-eye
Zircon

Sciatica

Amethyst
Chrysolite
Chrysoprase
Citrine
Diamond
Heliotrope
Jasper
Carnelian
Lapis lazuli
Magnetite
Malachite
Pearl
Pyrite
Rhodochrosite
Rose quartz
Ruby
Emerald
Tiger's-eye
Turquoise

Toenails

Amber
Chrysoprase
Diamond
Garnet
Island spar
Jasper
Carnelian
Onyx
Orange calcite
Pearl
Rhodochrosite
Rose quartz
Ruby
Sapphire
Turquoise
red Tourmaline

Blisters

Agate
Beryl
Indigolite
Rhodochrosite
Vanadinite
Aventurine
Diamond
Lapis lazuli
Turquoise
Zircon

Varicose veins

Aquamarine
Rock crystal
Amber
Chrysoprase
Diamond
Garnet
Carnelian
Lapis lazuli
Pearl
Ruby
Sapphire
layered Agate
Topaz
Zircon

254

Names and Terms

The following names and terms are explained. They belong to the special, and also wider, area of the gemstone studies/mineralogy, as well as to related sciences in which lithotherapy plays an important role.

Apache tear A small, round obsidian-aggregate, sometimes with a white rind, New Mexico, Texas.

Ammonites Extinct group of cuttlefish rich in forms, with chambered shells rolled in spirals. They are as fossils widespread all over the world.

Amulett stone Synonym for the Australian thunder egg with angular quartz filling.

Belemnites Extinct, resembling cuttlefish. In the vernacular, their "petrified" drawn-out, cone-shaped calcium shells are called thunderbolts or simply belemnites.

Bezoarstone Stonelike, globular form in the stomachs of mammals. It consists of hairs that are ingested by licking the animal pelt.

Biotite Dark, scaly mineral from the group of mica. Bleached, bronze-colored varieties are called Cats'-gold in the vernacular.

Bojistone Bulbous mineral aggregate (concretion) out of pyrite. Only place of discovery Kansas/USA.

Flint (Synonym: Firestone) Bulbous, roundish, Chalcedonic pebblestone. Mostly light gray to almost black, sometimes with a white rind.

Larimar Bluish to greenish white Pectolite variety from the Dominican Republic. On the market only since the 1970s. Very popular as a curative stone since it lends itself to easy cutting.

Islandspar (Synonym: Doublespar) Clear, colorless Calcite variety with pronounced double light-refraction.

Koprolite Fossilized excrement-ball of prehistoric living beings.

Meteoritestone (Synonym: Meteorstone, meteorite, aerolite) Solid fragment, come to the earth from the outer space, an outer-terrestrial stone.

Moqui-Marble Globular ironoxide-aggregate from Utah/USA.

Phantomquartz Rock crystal with layered inclusions that give the illusion of a zonal building.

Sceptrequartz Convex crystal on a small stem. Very common in all quartz varieties.

Amulettstones. Upper, from the left: Agate, Sugilite, Jasper. Below, from the left: Sodalite, Snowflake obsidian, Fluorite.

Bibliography

Anderson, B. W., and Jobbins, E. A. 1990. *Gem Testing*. Butterworths, London.

Arem, J. E. 1987. *Color Encyclopedia of Gemstones*. 2nd ed. Van Nostrand Reinhold Co., Inc., New York.

Bruton, E. 1977. *Diamonds*. 2nd ed. NAG, London.

CIBJO. 1975. *Bestimmungen zur Benennung und Beschreibung von Edelsteinen, Perlen, Kulturperlen, Synthesen, Imitationen*. Bern, Switzerland.

Copeland, L. L. 1960. *The Diamond Dictionary*. Los Angeles, CA, U.S.

Dake, H. C. 1950. *Northwest Gem Trails*. Portland, OR, U.S.

Fisher, P. J. 1966. *The Science of Gems*. Charles Scribner's Sons, New York.

Fleischer, M., and Mandarino, J. A. 1995. *Glossary of Mineral Species 1995*. The Mineralogical Record, Tucson, AZ, U.S.

Ford, W. E. 1932. *Dana's Textbook of Mineralogy*. John Wiley & Sons, New York.

Gill, J. O. 1978. *Gill's Index to Journals, Articles, and Books Relating to Gems and Jewelry*. Gemological Institute of America, Santa Monica, CA, U.S.

Green, T. 1981. *The World of Diamonds*. Weidenfield and Nicolson, London.

Henry, D. J. 1952. *Gem Trail Journal*. Long Beach, CA, U.S.

Jahns, R. H. 1975. "Gem Materials" in *Industrial Minerals and Rocks*. 4th ed. A.I.M.E., New York.

Kraus, E. H., and Slawson, C. B. 1947. *Gems and Gem Materials*. McGraw-Hill, New York.

Lewis, D. 1977. *Practical Gem Testing*. Northwood, London.

Liddicoat, R. T., Jr. 1989. *Handbook of Gem Identification*. Gemological Institute of America, Santa Monica, CA, U.S.

McIver, J. R. 1967. *Gems, Minerals, and Rocks in Southern Africa*. Elsevier (U.S.), New York.

Read, P. G. 1991 *Gemmology*. Butterworth-Heinemann Ltd., Oxford, UK.

Rouse, J. D. 1986. *Garnet*. Butterworths, London.

Shaub, B. M. 1975. *Treasures from the Earth*. Crown Publishers, New York.

Shipley, R. M. 1974. *Dictionary of Gems and Gemology*. Gemological Institute of America, Santa Monica, CA, U.S.

Sinkankas, J. 1970. *Gemstones of North America*. D. Van Nostrand, New York.

Sinkankas, J. 1972. *Gemstone and Mineral Data Book*. New York.

Smith, G. F. H., and Phillips, F. C. 1962. *Gemstones*. 13th ed. Methuen & Co., London.

U.S. Geological Survey (annual). Gemstones: ch. in *Minerals Yearbook*, U.S.G.S., Reston, VA, U.S.

Van Landingham, S. L. 1985. *Geology of World Gem Deposits*. Van Nostrand Reinhold Co., Inc., New York.

Webster, R., and Anderson, B. W. 1983. *Gems—Their Sources, Descriptions, and Identification*. Butterworths, London.

Yaverbaum, L. H. (ed.). 1980. *Synthetic Gems Production Techniques*. Noyes Data Corp., Park Ridge, NJ, U.S.

Gems and Gemology, Gemological Institute of America. Los Angeles, CA, U.S.

Gems and Minerals, Mentone, CA, U.S.

Journal of Gemmology, Gemmological Institute of Great Britain, London.

Journal of the German Gemmological Association. Idar-Oberstein, Germany.

Lapidary Journal. San Diego, CA, U.S.

The Mineralogical Record. Tucson, AZ, U.S.

Schweizer Strahler. Journal of the Swiss Association of Mineral Collectors and Polishers.

Illustration Acknowledgements

Photos: Dr. H. Bank, Idar-Oberstein: 56 top right; G. Becker, Idar-Oberstein: 89; E. A. Bunzel, Idar-Oberstein: 56 top left and below, 57 top and below left, 60 below, 61 below; Chudoba-Guebelin, *Edelsteinkundliches Handbuch*, Wilhelm Stolifuss Verlag, Bonn: 39; De Beers Consolidated Mines Ltd, Johannesburg, South Africa: 40, 50, 53, 59, 60 top, 72 left, 73, 75; H. Eisenbeiss, Munich: 34; Dr. E. Geubelin Lucerne: 52 below, 57 below right, 85, 93; K. Hartmann, Sobernheim: all gemstone plates as well as 29, 43, 47, 67; Her Majesty's Stationery Office, London: 9; Jain Cultured Pearls, *see* Heim/Bergstrasse: 61, 226, 227, 228; E. Pauly, Veitsrudt: 144; J. Petsch Jr., Idar-Oberstein: 52 top; A. Ruppenthal KG, Idar-Oberstein: 138, 140.
Drawings: H. Hoffmann, Munich, all diagrams apart from 17; W. Schumann: *Das Grosse Buch der Erde* (The Large Book of the Earth), Deutscher Buercherbund, Stuttgart: 17.
Chart: R. Webster: *The Gemmologists' Compendium*, N.A.G. Press Ltd, London.

Table of Constants

How to Use the Table

This table has been compiled to aid both the professional and the amateur to identify individual gemstones. It is designed to be used in conjunction with the standard gemological tests. It is not totally comprehensive, but it should enable the tester to eliminate some possibilities and perhaps suggest some, if he or she is faced with one of the more unusual examples.

Suppose that you have an unknown yellow gemstone to identify. Using the Table of Constants, this is how you go about identifying it:

1 Turn to pages 254–255, which cover yellow, orange, and brown stones.

2 Test for specific gravity (pages 23–25); this gives you a density of 3.65.

3 Test the refractive index (pages 31–33); this gives a reading of 1.738.

4 Run down the 1.700–1.799 column and across the 3.50–3.99 line until they meet. This gives you several possibilities.

5 Test for double refraction (page 34). If there is none, the field is narrowed to periclase or two members of the garnet group (pyrope and grossularite).

6 Turning to the garnet section in the main text, the tester will observe that pyrope and grossularite have distinctive absorption spectra. Using the spectroscope (pages 37–39), you can identify the gem if it shows one of these spectra.

Refr. index / Density	1.400–1.499	1.500	1.600–1.699
1.00–1.99	Ulexite 2–2½ Gaylussite 2½–3 Kurnakovite 3 Opal 5½–6½	Ulexite 2–2½ Amber 2–2½ Amber 2–3 Gaylussite 2½–3	
2.00–2.49	Ulexite 2 Natrolite 5–5½ Obsidian 5–5½ Haüynite 5½–6 Sodalite 5½–6 Opal 5½–6½	Meerschaum 2–2½ Amber 2–3 Colemanite 4½ Apophyllite 4½–5 Haüynite 5½–6 Leucite 5½–6 Petalite 6–6½ Hambergite 7½	Howlite 3–3½ Colemanite 4½ Hambergite 7½
2.50–2.99	Calcite 3 Coral 3–4 Onyx Marble 3½–4 Obsidian 5–5½ Haüynite 5½–6 Opal 5½–6½	Vivianite 1½–2 Pearl 2½–4½ Calcite 3 Howlite 3–3½ Coral 3–4 Anhydrite 3½ Argonite 3½–4 Dolomite 3½–4 Augelite 4½–5 Beryllonite 5½–6 Leucite 5½–6 Scapolite 5½–6 Sanidine 6 Labradorite 6–6½ Moonstone 6–6½ Rock Crystal 7 Smoky Quartz 7 Precious Beryl 7½–8	Vivianite 1½–2 Pearl 2½–4½ Calcite 3 Howlite 3–3½ Coral 3–4 Anhydrite 3½ Argonite 3½–4 Dolomite 3½–4 Onyx Marble 3½–4 Datolite 5–5½ Nephrite 6–6½ Danburite 7–7½ Tourmaline 7–7½ Precious Beryl 7½–8 Phenakite 7½–8
3.00–3.49	Fluorite 4	Magnesite 3½–4½ Amblygonite 6	Magnesite 3½–4½ Apatite 5 Hemimorphite 5 Datolite 5–5½ Diopside 5–6 Enstatite 5½ Amblygonite 6 Nephrite 6–6½ Jadeite 6½–7 Danburite 7–7½ Tourmaline 7–7½ Dumortierite 7–8½ Euclase 7½
3.50–3.99			Celestite 3–3½ Hemimorphite 5 Willemite 5½ Topaz 8
4.00–4.99		Witherite 3–3½	Barite 3–3½ Celestite 3–3½ Witherite 3–3½ Willemite 5½

The numbers following the gemstone names refer to Mohs' hardness

Gem Color White + Colorless + Gray

Refr. index Density	1.700–1.799	1.800–1.899	1.900 and higher
1.00–1.99			
2.00–2.49			
2.50–2.99			
3.00–3.49	Magnesite 3½–4½ Diopside 5–6 Clinozoisite 6–7		
3.50–3.99	Kyanite 4–7 Periclase 5½–6 Benitoite 6–6½ Grossularite 6½–7½ Sapphire 9	Benitoite 6–6½ Zircon 6½–7½	Sphalerite 3½–4 Anastase 5½–5 Zircon 6½–7½ Diamond 10
4.00–4.99	Sapphire 9	Zircon 6½–7½ YAG 8	Sphalerite 3½–4 Linobate 5½ Zircon 6½–7½
5.00–5.99			Scheelite 4½–5 Hematite 5½–6½ Strontium Titanate 5–6 Cubic Zirconia 8½
6.00–6.99		Cerussite 3–3½	Phosgenite 2–3 Cerussite 3–3½ Scheelite 4½–5 Cassiterite 6–7
7.00 and higher			Cassiterite 6–7 GGG 6½

Refr. index / Density	1.400–1.499	1.500–1.599	1.600–1.699
1.00–1.99	Kurnakovite 3 Opal 5½–6½	Amber 2–2½ Kurnakovite 3 Opal 5½–6½	
2.00–2.49	Natrolite 5–5½ Cancrinite 5–6 Opal 5½–6½ Tugtupite 5½–6	Stichtite 1½–2½ Gypsum 2 Meerschaum 2–2½ Apophyllite 4½–5 Thomsonite 5–5½ Cancrinite 5–6 Tugtupite 5½–6 Opal 5½–6½ Petalite 6–6½	
2.50–2.99	Calcite 3 Coral 3–4 Cancrinite 5–6 Tugtupite 5½–6 Aragonite 3½–4 Opal 5½–6½	Pearl 2½–4½ Calcite 3 Coral 3–4 Anhydrite 3½ Aragonite 3½–4 Dolomite 3½–4 Apophyllite 4½ Cancrinite 5–6 Scapolite 5½–6 Tugtupite 5½–6 Opal 5½–6½ Aventurine Feldspar 6–6½ Petrified Wood 6½–7 Jasper 6½–7 Amethyst 7 Rose Quartz 7 Precious Beryl 7½–8	Pearl 2½–4½ Calcite 3 Coral 3–4 Anhydrite 3½ Aragonite 3½–4 Dolomite 3½–4 Nephrite 6–6½ Danburite 7–7½ Tourmaline 7–7½ Precious Beryl 7½–8 Phenakite 7½–8
3.00–3.49	Fluorite 4		Rhodochrosite 4 Apatite 5 Nephrite 6–6½ Kunzite 6½–7 Jadeite 6½–7 Danburite 7–7½ Tourmaline 7–7½ Dumortierite 7–8½ Andalusite 7½ Topaz 8 Rhodizite 8–8½
3.50–3.99			Celestite 3–3½ Siderite 3½–4½ Rhodochrosite 4 Willemite 5½ Topaz 8
4.00–4.99			Barite 3–3½ Celestite 3–3½ Smithsonite 5 Willemite 5½
5.00–5.99			

The numbers following the gemstone names refer to Mohs' hardness

Gem Color Red + Pink + Orange

Refr. index / Density	1.700–1.799	1.800–1.899	1.900 and higher
1.00–1.99			
2.00–2.49			
2.50–2.99			
3.00–3.49	Rhodochrosite 4 Rhondonite 5½–6½ Clinozoisite 6–7	Rhodochrosite 4 Purpurite 4–4½	Purpurite 4–4½
3.50–3.99	Siderite 3½–4½ Rhodochrosite 4 Willemite 5½ Rhondonite 5½–6½ Benitoite 6–6½ Almandite 6½–7½ Hessonite 6½–7½ Pyrope 6½–7½ Rhodolite 6½–7½ Spinel 8 Taafeite 8–8½ Alexandrite 8½ Chrysoberyl 8½ Ruby 9 Sapphire 9	Siderite 3½–4½ Rhodochrosite 4 Sphene 5½–5 Benitoite 6–6½ Almandite 6½–7½ Zircon 6½–7½	Sphalerite 3½–4 Sphene 5–5½ Anastase 5½–6 Zircon 6½–7½ Diamond 10
4.00–4.99	Smithsonite 5 Willemite 5½ Almandite 6½–7½ Spessartite 6½–7½ Gahnite 7½–8 Painite 7½–8 Ruby 9 Sapphire 9	Smithsonite 5 Almandite 6½–7½ Spessartite 6½–7½ Zircon 6½–7½ Gahnite 7½–8 Painite 7½–8	Sphalerite 3½–4 Rutile 6–6½ Zircon 6½–7½
5.00–5.99			Proustite 2½ Crocoite 2½–3 Cuprite 3½–4 Zincite 4–5 Scheelite 4½–5 Hematite 5½–6½ Tantalite 6–6½
6.00–6.99			Crocoite 2½–3 Wulfenite 3 Cuprite 3½–4 Scheelite 4½–5 Tantalite 6–6½
7.00 and higher			Wulfenite 3 Tantalite 6–6½

Refr. index / Density	1.400–1.499	1.500–1.599	1.600–1.699
1.00–1.99	Gaylussite 2½–3 Opal 5½–6½	Amber 2–2½ Ivory 2–3	Jet 2½–4
2.00–2.49	Natrolite 5–5½ Obsidian 5–5½ Moldavite 5½	Meerschaum 2–2½ Serpentine 2½–5½ Apophyllite 4½–5 Leucite 5½–6 Petalite 6–6½ Hambergite 7½	Hambergite 7½
2.50–2.99	Calcite 3 Onyx Marble 3½–4 Obsidian 5–5½ Opal 5½–6½	Pearl 2½–4½ Serpentine 2½–5½ Calcite 3 Aragonite 3½–4 Obsidian 5–5½ Beryllonite 5½–6 Scapolite 5½–6 Sanidine 6 Aventurine Feldspar 6–6½ Moonstone 6–6½ Orthoclase 6–6½ Tiger's-Eye 6½–7 Aventurine 7 Citrine 7 Smoky Quartz 7 Iolite 7–7½ Precious Beryl 7½–8	Pearl 2½–4½ Calcite 3 Aragonite 3½–4 Dolomite 3½–4 Onyx Marble 3½–4 Datolite 5–5½ Brazilianite 5½ Nephrite 6–6½ Prehnite 6–6½ Danburite 7–7½ Tourmaline 7–7½ Precious Beryl 7½–8 Phenakite 7½–8
3.00–3.49	Fluorite 4	Ekanite 4½–6½ Amblygonite 6	Rhodochrosite 4 Apatite 5 Diopside 5–6 Hypersthene 5–6 Enstatite 5½ Actinolite 5½–6 Amblygonite 6 Nephrite 6–6½ Axinite 6½–7 Hiddenite 6½–7 Jadeite 6½–7 Kornerupine 6½–7 Peridot 6½–7 Sinhalite 6½–7 Danburite 7–7½ Tourmaline 7–7½ Dumortierite 7–8½ Andalusite 7½ Topaz 8
3.50–3.99			Siderite 3½–4½ Rhodochrosite 4 Willemite 5½ Topaz 8
4.00–4.99		Witherite 3–3½	Barite 3–3½ Witherite 3–3½

The numbers following the gemstone names refer to Mohs' hardness

Gem Color Yellow + Orange + Brown

Density \ Refr. index	1.700–1.799	1.800–1.899	1.900 and higher
1.00–1.99			
2.00–2.49			Sulfur 1½–2½
2.50–2.99			
3.00–3.49	Rhodochrosite 4 Hypersthene 5–6 Epidote 6–7 Clinozoisite 6–7 Axinite 6½–7 Peridot 6½–7 Sinhalite 6½–7 Idocrase 6½	Rhodochrosite 4 Purpurite 4–4½	Purpurite 4–4½
3.50–3.99	Siderite 3½–4½ Rhodochrosite 4 Kyanite 4–7 Hypersthene 5–6 Willemite 5½ Periclase 5½–6 Epidote 6–7 Sinhalite 6½–7 Grossularite 6½–7½ Pyrope 6½–7½ Staurolite 7–7½ Spinel 8 Chrysoberyl 8½ Sapphire 9	Siderite 3½–4½ Rhodochrosite 4 Sphene 5–5½ Andradite 6½–7½ Zircon 6½–7½	Sphalerite 3½–4 Sphene 5–5½ Anatase 5½–6 Andradite 6½–7½ Zircon 6½–7½ Diamond 10
4.00–4.99	Willemite 5½ Spessartite 6½–7½ Sapphire 9	Andradite 6½–7½ Spessartite 6½–7½ Zircon 6½–7½	Sphalerite 3½–4 Rutile 6–6½ Zircon 6½–7½
5.00–5.99			Crocoite 2½–3 Zincite 4–5 Scheelite 5½–5 Hematite 5½–6½ Tantalite 6–6½ Cubic Zirconia 8½
6.00–6.99		Cerussite 3–3½	Crocoite 2½–3 Phosphogenite 2–3 Wulfenite 3 Cerussite 3–3½ Scheelite 4½–5 Tantalite 6–6½ Cassiterite 6–7
7.00 and higher			Wulfenite 3 Tantalite 6–6½ Cassiterite 6–7

264

Refr. index Density	1.400–1.499	1.500–1.599	1.600–1.699
1.00–1.99	Opal 5½–6½	Amber 2–2½	Turquoise 5–6
2.00–2.49	Chrysocolla 2–3 Obsidian 5–5½ Moldavite 5½ Haüynite 5½–6 Opal 5½–6½	Chrysocolla 2–3 Serpentine 2½–5½ Variscite 4–5 Apophyllite 4½–5 Obsidian 5–5½ Thomsonite 5–5½ Moldavite 5½ Haüynite 5½–6	Turquoise 5–6
2.50–2.99	Onyx Marble 3½–4 Obsidian 5–5½ Haüynite 5½–6 Opal 5½–6½	Vivianite 1½–2 Pearl 2½–4½ Serpentine 2½–5½ Variscite 4–5 Apophyllite 4½–5 Wardite 5 Obsidian 5–5½ Amazonite 6–6½ Chyroprase 6½–7 Aventurine 7 Prasiolite 7 Aquamarine 7½–8 Precious Beryl 7½–8 Emerald 7½–8	Vivianite 1½–2 Pearl 2½–4½ Onyx Marble 3½–4 Datolite 5–5½ Tremolite 5–6 Turquoise 5–6 Brazilianite 5½ Nephrite 6–6½ Prehnite 6–6½ Tourmaline 7–7½ Precious Beryl 7½–8 Emerald 7½–8
3.00–3.49	Fluorite 4	Ekanite 4½–6½	Malachite 3½–4 Apatite 5 Dioptase 5 Hemimorphite 5 Datolite 5–5½ Diopside 5–6 Hypersthene 5–6 Tremolite 5–6 Enstatite 5½ Smaragdite 5½ Actinolite 5½–6 Nephrite 6–6½ Hiddenite 6½–7 Jadeite 6½–7 Kornerupine 6½–7 Peridot 6½–7 Sinhalite 6½–7 Tourmaline 7–7½ Andalusite 7½ Euclase 7½
3.50–3.99			Celestite 3–3½ Malachite 3½–4 Hemimorphite 5 Hypersthene 5–6 Willemite 5½ Topaz 8
4.00–4.99			Barite 3–3½ Smithsonite 5

The numbers following the gemstone names refer to Mohs' hardness

Gem Color Green + Yellow-green + Blue-green

Refr. index / Density	1.700–1.799	1.800–1.899	1.900 and higher
1.00–1.99			
2.00–2.49			Sulfur 1½–2½
2.50–2.99			
3.00–3.49	Malachite 3½–4 Dioptase 5 Diopside 5–6 Hypersthene 5–6 Epidote 6–7 Clinozoisite 6–7 Idocrase 6½ Peridot 6½–7 Sinhalite 6½–7	Malachite 3½–4 Uvarovite 6½–7½	Malachite 3½–4
3.50–3.99	Malachite 3½–4 Kyanite 4–7 Hypersthene 5–6 Willemite 5½ Periclase 5½–6 Epidote 6–7 Sinhalite 6½–7 Grossularite 6½–7½ Spinel 8 Taaffeite 8–8½ Alexandrite 8½ Chrysoberyl 8½ Sapphire 9	Malachite 3½–4 Sphene 5–5½ Demantoid 6½–7½ Uvarovite 6½–7½ Zircon 6½–7½	Malachite 3½–4 Sphalerite 3½–4 Sphene 5–5½ Zircon 6½–7½ Diamond 10
4.00–4.99	Malachite 3½–4 Smithsonite 5 Willemite 5½ Gahnite 7½–8 Sapphire 9	Malachite 3½–4 Smithsonite 5 Zircon 6½–7½ Gahnite 7½–8	Malachite 3½–4 Sphalerite 3½–4 Zircon 6½–7½
5.00–5.99			Cubic Zirconia 8½
6.00–6.99			Phosgenite 2–3
7.00 and higher			

Refr. index / Density	1.400–1.499	1.500–1.599	1.600–1.699
1.00–1.99	Opal 5½–6½	Amber 2–2½ Opal 5½–6½	
2.00–2.49	Chrysocolla 2–4 Haüynite 5½–6 Sodalite 5½–6 Opal 5½–6½	Gypsum 2 Chrysocolla 2–4 Variscite 4–5 Apophyllite 4½–5 Haüynite 5½–6 Opal 5½–6½	Turquoise 5–6
2.50–2.99	Lapis lazuli 5–6 Haüynite 5½–6 Opal 5½–6½	Vivianite 1½–2 Pearl 2½–4½ Coral 3–4 Anhydrite 3½ Variscite 4–5 Apophyllite 4½–5 Wardite 5 Lapis lazuli 5–6 Haüynite 5½–6 Opal 5½–6 Amazonite 6–6½ Chalcedony 6½–7 Chrysoprase 6½–7 Jasper 6½–7 Aventurine 7 Prasiolite 7 Iolite 7–7½ Aquamarine 7½–8 Emerald 7½–8	Vivianite 1½–2 Pearl 2½–4½ Anhydrite 3½ Turquoise 5–6 Nephrite 6–6½ Tourmaline 7–7½
3.00–3.49	Fluorite 4 Lapis lazuli 5–6	Lapis lazuli 5–6	Apatite 5 Dioptase 5 Hemimorphite 5 Lazulite 5–6 Nephrite 6–6½ Clinozoisite 6–7½ Sillimanite 6–7½ Smaragdite 6½ Axinite 6½–7 Jadeite 6½–7 Tanzanite 6½–7 Tourmaline 7–7½ Dumortierite 7–8½ Euclase 7½ Topaz 8
3.50–3.99			Celestite 3–3½ Hemimorphite 5 Topaz 8
4.00–4.99			Barite 3–3½ Celestite 3–3½ Smithsonite 5

The numbers following the gemstone names refer to Mohs' hardness

Gem Color Blue + Blue-green + Blue-red

Refr. index / Density	1.700–1.799	1.800–1.899	1.900 and higher
1.00–1.99			
2.00–2.49			
2.50–2.99			
3.00–3.49	Dioptase 5 Idocrase 6½ Axinite 6½–7 Tanzanite 6½–7	Purpurite 4–4½	Purpurite 4–4½
3.50–3.99	Azurite 3½–4 Kyanite 4½–7 Benitoite 6–6½ Spinel 8 Taaffeite 8–8½ Ruby 9 Sapphire 9	Azurite 3½–4 Benitoite 6–6½ Zircon 6½–7½	Anatase 5½–6 Zircon 6½–7½ Diamond 10
4.00–4.99	Smithsonite 5 Gahnite 7½–8 Ruby 9 Sapphire 9	Smithsonite 5 Zircon 6½–7½ Gahnite 7½–8	Zircon 6½–7½
5.00–5.99			

Refr. index / Density	1.400–1.499	1.500–1.599	1.600–1.699
1.00–1.99	Opal 5½–6½	Amber 2–2½ Opal 5½–6½	
2.00–2.49	Tugtupite 5½–6 Opal 5½–6½	Stichtite 1½–2½ Tugtupite 5½–6 Opal 5½–6½	
2.50–2.99	Calcite 3 Coral 3–4 Lapis lazuli 5–6 Tugtupite 5½–6 Opal 5½–6½	Calcite 3 Coral 3–4 Anhydrite 3½ Charoite 5–6 Lapis lazuli 5–6 Scapolite 5½–6 Tugtupite 5½–6 Opal 5½–6½ Petrified Wood 6½–7 Jasper 6½–7 Amethyst 7 Amethyst Quartz 7 Rose Quartz 7 Iolite 7–7½	Calcite 3 Coral 3–4 Anhydrite 3½ Nephrite 6–6½ Sugilite 6–6½ Tourmaline 7–7½
3.00–3.49	Fluorite 4 Lapis lazuli 5–6	Lapis lazuli 5–6 Amblygonite 6	Apatite 5 Amblygonite 6 Nephrite 6–6½ Axinite 6½–7 Jadeite 6½–7 Kunzite 6½–7 Tanzanite 6½–7 Dumortierite 7–8½ Tourmaline 7–7½ Topaz 8
3.50–3.99			Topaz 8
4.00–4.99			Smithsonite 5
5.00–5.99			
6.00–6.99			
7.00 and higher			

The numbers following the gemstone names refer to Mohs' hardness

Gem Color Violet + Blue-red

Refr. index / Density	1.700–1.799	1.800–1.899	1.900 and higher
1.00–1.99			
2.00–2.49			
2.50–2.99			
3.00–3.49	Axinite 6½–7 Tanzanite 6½–7	Purpurite 4–4½	Purpurite 4–4½
3.50–3.99	Benitoite 6–6½ Almandite 6½–7½ Spinel 8 Taaffeite 8–8½ Ruby 9 Sapphire 9	Benitoite 6–6½ Almandite 6½–7½ Zircon 6½–7½	Zircon 6½–7½
4.00–4.99	Smithsonite 5 Almandite 6½–7½ Gahnite 7½–8 Ruby 9 Sapphire 9	Smithsonite 5 Almandite 6½–7½ Zircon 6½–7½ Gahnite 7½–8	Zircon 6½–7½
5.00–5.99			Proustite 2½ Cuprite 3½–4 Tantalite 6–6½ Cubic Zirconia 8½
6.00–6.99			Cuprite 3½–4 Tantalite 6–6½
7.00 and higher			Tantalite 6–6½

Refr. index / Density	1.400–1.499	1.500–1.599	1.600–1.699
1.00–1.99	Opal 5½–6½	Amber 2–2½ Coral 3–4 Opal 5½–6½	Jet 2½–4
2.00–2.49	Obsidian 5–5½ Sodalite 5½–6 Opal 5½–6½	Meerschaum 2–2½ Obsidian 5–5½ Opal 5½–6½ Hambergite 7½	Hambergite 7½
2.50–2.99	Obsidian 5–5½ Opal 5½–6½	Pearl 2½–4½ Aragonite 3½–4 Obsidian 5–5½ Opal 5½–6½ Sanidine 6 Labradorite 6–6½ Chalcedony 6½–7 Petrified Wood 6½–7 Jasper 6½–7 Smoky Quartz 7	Pearl 2½–4½ Aragonite 3½–4 Nephrite 6–6½ Tourmaline 7–7½
3.00–3.49	Fluorite 4		Hypersthene 5–6 Nephrite 6–6½ Jadeite 6½–7 Tourmaline 7–7½
3.50–3.99			Hypersthene 5–6
4.00–4.99			
5.00–5.99			
6.00–6.99			
7.00 and higher			

The numbers following the gemstone names refer to Mohs' hardness

Gem Color Black + Gray

Refr. index / Density	1.700–1.799	1.800–1.899	1.900 and higher
1.00–1.99			
2.00–2.49			
2.50–2.99			
3.00–3.49	Hypersthene 5–6 Epidote 6–7		
3.50–3.99	Hypersthene 5–6 Epidote 6–7 Staurolite 7–7½ Spinel 8 Sapphire 9	Andradite 6½–7½	Anatase 5½–6 Andradite 6½–7½ Diamond 10
4.00–4.99	Gahnite 7½–8 Sapphire 9	Andradite 6½–7½ Gahnite 7½–8	Chromite 5½ Rutile 6–6½ Andradite 6½–7½
5.00–5.99			Hematite 5½–6½
6.00–6.99		Cerussite 3–3½	Cerussite 3–3½
7.00 and higher			

Refr. index / Density	1.400–1.499	1.500–1.599	1.600–1.699
1.00–1.99	Opal 5½–6½	Opal 5½–6½	
2.00–2.49	Opal 5½–6½ Tugtupite 5½–6	Serpentine 2½–5½ Howlite 3–3½ Tugtupite 5½–6 Opal 5½–6½	Howlite 3–3½ Turquoise 5–6
2.50–2.99	Onyx Marble 3½–4 Lapis lazuli 5–6 Tugtupite 5½–6 Opal 5½–6½	Serpentine 2½–5½ Howlite 3–3½ Onyx Marble 3½–4 Tufa 3½–4 Ammonite 4 Charoite 5–6 Lapis lazuli 5–6 Tugtupite 5½–6 Opal 5½–6½ Jade-Albite 6 Aventurine 6–6½ Labradorite 6–6½ Moonstone 6–6½ Peristerite 6–6½ Agate 6½–7 Chalcedony 6½–7 Petrified Wood 6½–7 Jasper 6½–7 Moss Agate 6½–7 Amethyst Quartz 7 Aventurine 7 Tiger's-Eye 7	Howlite 3–3½ Onyx Marble 3½–4 Tufa 3½–4 Ammonite 4 Turquoise 5–6 Nephrite 6–6½ Tourmaline 7–7½
3.00–3.49	Fluorite 4 Lapis lazuli 5–6	Lapis lazuli 5–6	Malachite 3½–4 Rhodochrosite 4 Nephrite 6–6½ Jadeite 6½–7 Tourmaline 7–7½
3.50–3.99			Malachite 3½–4 Rhodochrosite 4
4.00–4.99			Malachite 3½–4
5.00–5.99			
6.00–6.99			
7.00 and higher			

The numbers following the gemstone names refer to Mohs' hardness

Gem Color Multicolored + Phenomenal

Refr. index / Density	1.700–1.799	1.800–1.899	1.900 and higher
1.00–1.99			
2.00–2.49			
2.50–2.99			
3.00–3.49	Malachite 3½–4 Rhodochrosite 4 Rhodonite 5½–6	Malachite 3½–4 Rhodochrosite 4	Malachite 3½–4
3.50–3.99	Malachite 3½–4 Rhodochrosite 4 Rhodonite 5½–6½ Alexandrite 8½	Malachite 3½–4 Rhodochrosite 4	Malachite 3½–4
5.00–5.99			
6.00–6.99			
7.00 and higher			

Index

Page numbers printed in bold indicate the main reference.

Photo on page 1 shows golden beryl.
Photo on pages 2–3 and front cover shows rhodochrosite, enlarged 3 times.
Photo on pages 4–5 shows tourmaline, enlarged 4 times.
Photo on page 6 shows tourmaline, slightly enlarged.

Library of Congress Cataloging-in-Publication Data

Schumann, Walter,
  [Edelsteine und Schmucksteine. English]
  Gemstones of the world / Walter Schumann.—Rev. & expanded ed.
    p. cm.
  Includes bibliographical references (p. –) and index.
  ISBN 0-8069-9461-4
  1. Precious stones I. Title.
  QE392.S54513   1997
  553.8–dc21

                           97–380
                           CIP

Translated by Annette Englander

10 9 8 7 6 5 4 3

Published 1997 by Sterling Publishing Company, Inc.
387 Park Avenue South, New York, N.Y. 10016
Originally published by BLV Verlagsgesellschaft mbH
under the title *Edelsteine und Schmucksteine*
© 1999 by BLV Verlagsgesellschaft mbH, Munich
English translation © 1997, 1999 by Sterling Publishing Co., Inc.
Distributed in Canada by Sterling Publishing
⅓ Canadian Manda Group, 165 Dufferin Street
Toronto, Ontario, Canada M6K 3H6
Distributed in Australia by Capricorn Link (Australia) Pty. Ltd.
P.O. Box 704, Windsor, NSW 2756, Australia
Printed in Germany
All rights reserved

Sterling ISBN 0-8069-0028-8

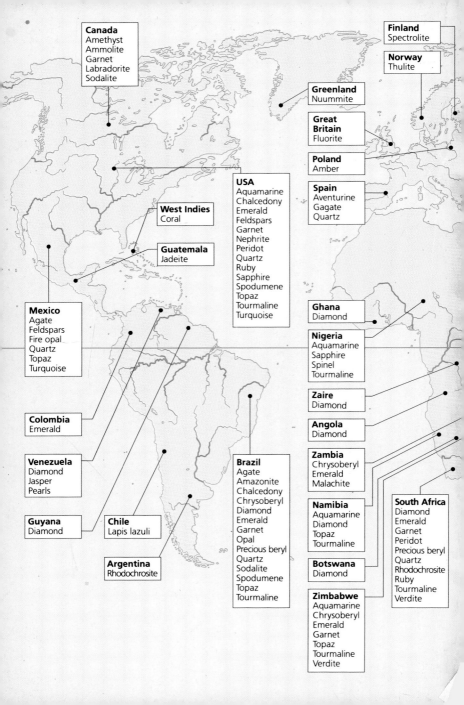

Canada
Amethyst
Ammolite
Garnet
Labradorite
Sodalite

Finland
Spectrolite

Norway
Thulite

Greenland
Nuummite

Great Britain
Fluorite

Poland
Amber

USA
Aquamarine
Chalcedony
Emerald
Feldspars
Garnet
Nephrite
Peridot
Quartz
Ruby
Sapphire
Spodumene
Topaz
Tourmaline
Turquoise

Spain
Aventurine
Gagate
Quartz

West Indies
Coral

Guatemala
Jadeite

Mexico
Agate
Feldspars
Fire opal
Quartz
Topaz
Turquoise

Ghana
Diamond

Nigeria
Aquamarine
Sapphire
Spinel
Tourmaline

Zaire
Diamond

Angola
Diamond

Colombia
Emerald

Venezuela
Diamond
Jasper
Pearls

Brazil
Agate
Amazonite
Chalcedony
Chrysoberyl
Diamond
Emerald
Garnet
Opal
Precious beryl
Quartz
Sodalite
Spodumene
Topaz
Tourmaline

Zambia
Chrysoberyl
Emerald
Malachite

South Africa
Diamond
Emerald
Garnet
Peridot
Precious beryl
Quartz
Rhodochrosite
Ruby
Tourmaline
Verdite

Guyana
Diamond

Chile
Lapis lazuli

Namibia
Aquamarine
Diamond
Topaz
Tourmaline

Botswana
Diamond

Argentina
Rhodochrosite

Zimbabwe
Aquamarine
Chrysoberyl
Emerald
Garnet
Topaz
Tourmaline
Verdite